Can a Darwinian Be a Christian?

The Relationship between Science and Religion

This book addresses a question at the heart of the current debate about the relationship between science and religion, in particular between that form of evolutionary biology known as Darwinism and the basic tenets of the Christian faith. The question is: Can someone who accepts Darwin's theory of natural selection subscribe at the same time to the essential claims of Christianity?

Adopting a balanced perspective on the subject, Michael Ruse offers a serious examination of both Darwinism and Christianity. He covers a wide range of topics from the Scopes Monkey Trial to claims about the religious significance of extraterrestrials. He deals with major figures in the current science/religion debate (Richard Dawkins, Stephen Jay Gould, and E. O. Wilson on the science side, as well as Arthur Peacocke, Robert J. Russell, and Keith Ward on the religion side). He considers in detail the claims of the new creationism and reveals some surprising parallels between Darwinian materialists and traditional thinkers such as Saint Augustine.

Michael Ruse argues that, although it is at times difficult for a Darwinian to embrace Christian belief, it is by no means inconceivable. At the same time he suggests ways in which a Christian believer should have no difficulty accepting evolution in general, and Darwinism in particular.

Writing with verve and avoiding technical jargon, Michael Ruse has produced an important contribution to a sometimes overheated debate for anyone interested in, and perhaps even troubled by, these issues who seeks an informed and judicious guide.

The author of many books on Darwin and evolutionary biology, Michael Ruse is Lucyle T. Werkmeister Professor of Philosophy at the Florida State University. He has been much involved in debates with creationists and was an expert witness for the ACLU in the 1981 Arkansas creation trial when he spoke to the questions of the philosophy of evolution as opposed to that of creationism.

Can a Darwinian Be a Christian?

The Relationship between Science and Religion

MICHAEL RUSE

Florida State University

CAMBRIDGE
UNIVERSITY PRESS

CAMBRIDGE UNIVERSITY PRESS
Cambridge, New York, Melbourne, Madrid, Cape Town, Singapore, São Paulo

Cambridge University Press
40 West 20th Street, New York, NY 10011–4211, USA
www.cambridge.org
Information on this title:www.cambridge.org/9780521631440

First published 2000
Reprinted 2001 (thrice), 2002
First paperback edition published 2004
Reprinted 2004, 2005, 2006

Printed in the United States of America

A catalogue record for this book is available from the British Library.

Library of Congress Cataloguing in Publication Data
Ruse, Michael.
Can a Darwinian be a Christian? : the relationship between science and religion /
Michael Ruse.
p.   cm.
Includes bibliographical references and index.
ISBN 0-521-63144-0
1. Evolution (Biology) — Religious aspects — Christianity.   2. Evolution
(Biology) — Philosophy.   3. Darwin, Charles, 1809-1882.   4. Religion and sciences.
I. Title.
BT712 R87 2000
231.7'652—dc21   99-462245

ISBN-13   978-0-521-63144-0 hardback
ISBN-10   0-521-63144-0 hardback

ISBN-13   978-0-521-63716-9 paperback
ISBN-10   0-521-63716-3 paperback

To the memory of my parents,
William and Margaret Ruse

Contents

Preface

Let me be open. I think that evolution is a fact and that Darwinism rules triumphant. Natural selection is not simply an important mechanism. It is the only significant cause of permanent organic change. I stand somewhere to the right of Archdeacon Paley on adaptation and design. I see purpose and function everywhere. I am an ardent naturalist and an enthusiastic reductionist, and those who disagree with me are wimps. I think that everything applies to humans, thought and action, and that sociobiology is the best thing to happen to the social sciences in the last century. The kindest thing that can be said for those who disagree – Marxists, feminists, constructivists, and fellow travellers   is that they speak from ignorance. Perhaps their genes make them do it.

Yet, all of this said, I cannot for the life of me see why so many – Darwinians and Christians alike – think that such a position as mine implies an immediate and emphatically negative response to the question I have posed in my title. Why should the devil have all the good tunes? Why should the devil have all the good science and philosophy? Saint Augustine and Saint Thomas Aquinas would be appalled at such a presumption, and we should feel the same way. It may indeed be the case that a Darwinian cannot be a Christian, but this is something to be decided only after one has looked at the two systems and worked through their points of possible conflict and dispute. It is not to be settled a priori before one begins. It is certainly not to be settled in happy and total ignorance of what others claim and believe.

Because I feel so strongly about this, I have decided to take seriously my own admonition. What you have before you are the fruits of my labours. I will say that I have been surprised at some of the things I have learnt and that I now hold some conclusions that I would hitherto have rejected. But I should also say that I do not much care whether you or anyone else end by agreeing or disagreeing with me. I do care that you think my inquiry is serious and worth making and that my arguments, especially if you disagree with them, are worth considering. What I will say is that I have had more fun on this project than on anything similar for many years. It will be reward enough, if I can infect you with my enthusiasm and the sheer joy of intellectual inquiry, comparing and contrasting two major world systems. What more could either a Darwinian or a Christian ask of life?

I am much in the debt of many people who have listened to me, supplied me with references, and explained to me points of Christian belief. These include Francisco J. Ayala, Philip Hefner, Ernan McMullin, Arthur Peacocke, Robert J. Russell, and Keith Ward. For technical scientific advice, I am indebted to Ursula Goodenough, Russell Doolittle, and Kenneth Miller. Michael Behe, William Dembski, Phillip Johnson, Alvin Plantinga, and all of their fellow thinkers whom I criticize strongly have been unfailingly courteous and friendly, showing that one can have major intellectual differences which need not (and should not) translate into personal attitudes or behaviour. My fellow evolutionists Richard Dawkins, Daniel Dennett, Stephen Jay Gould, William Provine, and Edward O. Wilson have long shown that our complete disagreement on the science–religion relationship is far less important than the quest for truth and the warmth of friendship. Four people very kindly read the whole manuscript and gave me advice: John Beatty, Eduardo R. Cruz, Edward Oakes, S. J., and Robert Pennock. I am particularly indebted to the history and philosophy of science graduate students at the University of Minnesota who put me through a gruelling examination on the contents of the work. My biblical references are to the Revised Standard Version, and I feel a traitor to the King James Version. For capitalization of religious terms, and in making references to deity, I have followed the recommended practice of the *Chicago Manual of Style,* fourteenth edition. My editor at Cambridge University Press, Terence Moore, deserves many thanks for many things, not the least of which was suggesting the topic in the first place. My assistant, Alan Belk, and my secretary, Linda Jenkins, worked beyond the call of duty. As always, my family – my wife

Lizzie and our children Emily, Oliver, and Edward – gave support and love.

One final word. I was born in Birmingham in the British Midlands in 1940, at the beginning of the Second World War. My father was a conscientious objector, and this brought him into contact with members of the Religious Society of Friends, the Quakers. After the war, he and my mother joined the society, and it was within this group that my sister and I were raised until 1953, when our mother died and the family that had been was no more. That was all long ago and far away, but every day I am aware that the deepest influence on my life was that loving Christian atmosphere created by my parents and their coreligionists in the Warwickshire Monthly Meeting, with which our local group was affiliated. If any of my readers feel that there is something in these pages which helps them on the spiritual route that we all must travel, thank those very ordinary and very wonderful people, not me.

Prologue

"I should like to ask Professor Huxley, who is sitting by me, and is about to tear me to pieces when I have sat down, as to his belief in being descended from an ape. Is it on his grandfather's or his grandmother's side that the ape ancestry comes in?" And then taking a graver tone, he asserted, in a solemn peroration, that Darwin's views were contrary to the revelation of God in the Scriptures. Professor Huxley was unwilling to respond: but he was called for, and spoke with his usual incisiveness and with some scorn: "I am here only in the interests of science," he said, "and I have not heard anything which can prejudice the case of my august client." Then after showing how little competent the Bishop was to enter upon the discussion, he touched on the question of Creation. "You say that development drives out the Creator; but you assert that God made you: and yet you know that you yourself were originally a little piece of matter, no bigger than the end of this gold pencil-case." Lastly as to the descent from a monkey, he said: "I should feel it no shame to have risen from such an origin; but I should feel it a shame to have sprung from one who prostituted the gifts of culture and eloquence to the service of prejudice and of falsehood." (Huxley 1900, 1, 200–201)

Stirring stuff. Samuel Wilberforce, bishop of Oxford, member of the House of Lords, leader of the Church of England, clashes with Thomas Henry Huxley, the sometime scholarship boy and then naval surgeon, now a morphologist and paleontologist and professor at the London School of Mines – a place whose educational rank with respect to Oxbridge may be likened to the status of the Salvation Army with respect to the Anglican Church. The champion of religion and the authority of the Bible, the leader of the "high church" movement, takes on the defender of science, the "bulldog," speaking for the new theory of organic evolu-

1

tion. And the spokesman for tradition and power comes away with his nose well and truly bloodied!

I still remember, some forty years ago, my history master – a good old-fashioned rationalist he – keeping us enthralled with his reenactment of the great battle. The smarmy episcopal debating tricks shown mere tinsel and illusion by the cold hard logic of the man of integrity and science. It is little wonder that the Wilberforce/Huxley fight, at the annual meeting of the British Association for the Advancement of Science in Oxford in 1860, one year after Charles Darwin had published his *On the Origin of Species,* has become the stuff of legend. In the history of science it ranks right up there with Archimedes stepping into his bath and with the aged Galileo down on his knees, recanting his Copernicanism while defiantly whispering: "But it does move, after all!"

It is a grand story, this public clash between the titans of Church and Science back at the height of the Victorian era. Yet, there are also tales from the century just gone. Move the clock forward to the 1920s, and cross the Atlantic to the courthouse in Dayton, Tennessee. It is a hot day, so hot in fact that the judge has moved everyone outside: defendant John Thomas Scopes, a local schoolteacher accused of teaching Darwinian evolutionary theory in violation of state law; prosecuting attorney and self-proclaimed expert witness on the Bible, William Jennings Bryan, spell-binding orator and three-time former presidential candidate; and Clarence Darrow, deadly effective defence attorney and notorious agnostic.

Darrow picked up the Bible and began to read: "'And the Lord God said unto the serpent, Because thou hast done this, thou art cursed above all cattle, and above every beast of the field; upon thy belly shalt thou go and dust shalt thou eat all the days of thy life.' Do you think that is why the serpent is compelled to crawl upon its belly?"

"I believe that."

"Have you any idea how the snake went before that time?"

"No sir."

"Do you know whether he walked on his tail or not?"

"No, sir, I have no way to know."

There was a howl of laughter from the crowd.

Suddenly Bryan's voice rose, screaming, hysterical: "The only purpose Mr. Darrow has is to slur at the Bible. . . . I want the world to know that this man, who does not believe in a God, is trying to use a court in Tennessee – "

"I object to your statement." Darrow was contemptuous. "I am examining you on your fool ideas that no intelligent Christian on earth believes."

Judge Raulston put an end to the argument by adjourning the court.

That night, at last, it rained. (Settle 1972, 108–9)

A wonderful story and, very thinly disguised, a wonderful film: *Inherit the Wind.* The Darrow figure (played by Spencer Tracey) vanquishes the Bryan figure (played by Frederick March), yet shows tolerance and understanding when, left alone at the end in the courtroom, he picks up the Bible and Darwin and thrusts both into his carrying bag. It is true that Scopes was found guilty and fined $100, although as a matter of fact this was overturned on a technicality on appeal. It is also true that the Tennessee law remained on the books for another forty years. But evolution had triumphed and Christianity had lost. Thanks particularly to the vitriolically funny reporting of H. L. Mencken of the *Baltimore Sun*, the world laughed at Tennessee and its antiquated beliefs: "degraded nonsense which country preachers are ramming and hammering into yokel skulls" (Ginger 1958, 129).

A third and final vignette. In 1957, during the depths of the cold war, the Russians scored a stunning triumph of technology when they put aloft a satellite known as sputnik. Even more than for what it was, the Russian achievement counted for what it was perceived to be. Above all, it became a triumph of propaganda. Reaching back to when I was a schoolboy, I can remember when the Russians described their second satellite as being so large that it was equal in size to that epitome of American success and opulence, a Cadillac. Desperately trying to regain ground, America poured money into science and technology. Included in that effort was science education, which in turn led to a series of wonderful textbooks, the biological representatives of which had full and detailed and enthusiastic discussions of evolution and its causes. At once conservative Christians swung into action, opposing evolution and promoting an alternative which became known as "creation science," something suspiciously like the early chapters of Genesis taken absolutely literally. And so, eventually, in 1981 in the state of Arkansas, there was another court trial, at which the American Civil Liberties Union – that organization determined to defend the integrity and authority of the U. S. Constitution – brought suit against a new law which insisted on the teaching of creation science alongside evolution. Among the expert witnesses for the plaintiffs was a professor from Canada.

Q: Dr. Ruse, having examined the creationist literature at great length, do you have a professional opinion about whether creation science measures up to the standards and characteristics of science that you have just been describing?

A: Yes, I do. In my opinion, creation science does not have those attributes that distinguish science from other endeavours.

Q: Would you please explain why you think it does not.

A: Most importantly, creation science necessarily looks to the supernatural acts of a Creator. According to creation-science theory, the Creator has intervened in supernatural ways using supernatural forces.

Q: Do you think that creation science is testable?

A: Creation science is neither testable nor tentative. Indeed, an attribute of creation science that distinguishes it quite clearly from science is that it is absolutely certain about all of the answers. And considering the magnitude of the questions it addresses – the origins of man, life, the earth, and the universe – that certainty is all the more revealing. Whatever the contrary evidence, creation science never accepts that its theory is falsified. This is just the opposite of tentativeness and makes a mockery of testing.

Q: Do you find that creation science measures up to the methodological considerations of science?

A: Creation science is woefully lacking in this regard. Most regrettably, I have found innumerable instances of outright dishonesty, deception, and distortion used to advance creation-science arguments.

Q: Dr. Ruse, do you have an opinion to a reasonable degree of professional certainty about whether creation science is science?

A: Yes.

Q: What is your opinion?

A: In my opinion creation science is not science.

Q: What do you think it is?

A: As someone also trained in the philosophy of religion, in my opinion creation science is religion. (Ruse 1988a, 304–6)

Terrific testimony! Modesty must not stop me from putting myself in line: Thomas Henry Huxley, Clarence Darrow, and now Michael Ruse. Little wonder that the judge found definitively in favour of evolution, throwing creation science out of court and out of classroom with the enthusiasm and effectiveness of a bruiser at a nightclub. So much for the religious opposition to evolution. Three strikes and you're out!

Would that life sometimes coincided with legend. I am sorry to have to tell you that – probably as with Archimedes and Galileo both – not one of the three stories I have just told you is true. At least (for I do not want to make myself into a liar before I begin), not one of the stories is the triumph of light over darkness that popular history would have it. Take the Wilberforce/Huxley clash. For a start, there was certainly not the dramatic tension and stark opposition conveyed by the passage which I quoted at the beginning of this Prologue – a passage, incidentally, taken from Huxley's *Life and Letters* as compiled by a dutiful son. Reports from the time suggest that everybody enjoyed himself immensely, and all went cheerfully off to dinner together afterwards. It was only later that the encounter was made into the ultimate fight for the soul of science, for the disinterested pursuit of truth no matter what the cost and what the opposition of vested interests (Lucas 1979; Jensen 1991).

And as with all legends, from Abraham to Princess Diana, there was a reason why this one took on the form and aura that it did. The second half of the nineteenth century, in Britain and elsewhere (Germany and America, particularly), was the time when the educated middle class were moving to make a full and recognized place for themselves in the new society, an urban industrialized society replacing the rural near-feudalism of the past. They wanted a civil service which was based on merit and not privilege; they wanted similar reformation of the professions and the military; they wanted education for all from the earliest years, and they wanted it to be secular, no longer under the thrall of an established church; they wanted respect for science and related areas like medicine; and they wanted the opportunity to make a lifetime career without need of inherited income or a rich patron. They wanted . . . They wanted all of these things and more, and were prepared to fight hard to realize their ends (Ruse 1996a). It is little wonder that they sought and found their myths and legends. As the tale of Moses leading his people from Egypt has bound together and inspired generations of Jews, as the story of Drake and the Spanish Armada has filled countless Britons with pride against the Catholic hordes of the continent, so the telling of Huxley's destruction of the pompous prelate has drawn many a bright schoolchild away from the classics and into a life of science.

The Scopes Monkey Trial has assumed no less mythic proportions: the very name "Scopes Monkey Trial" should alert you to that. And again, as we start to dig into the real details its iconic status as a clash between science and religion, between Evolution and Christianity, starts to seem rather less secure. For a start, it seems highly probable that the good citizens of Dayton, Tennessee, themselves brought on the trial, not because they were appalled at the teaching of evolution – they really did not care too much about that – but because they thought the publicity would be good for business. Scopes himself was a respected and loved teacher, who allowed himself to be put on the firing line and prosecuted – even though there is some doubt as to whether he ever actually taught evolution in class at all! Although the judge did not permit them to testify in court, ready to speak on Scopes's behalf were four ordained clergymen, including Shailer Mathews, dean of the Divinity School at the University of Chicago. William Jennings Bryan was anomalous, not only in being prepared to act as expert witness as well as legal counsel for the prosecution but also in openly rejecting literal interpretations of the Bible. He quipped that he took his stand on the Rock of Ages rather than the age of rocks, but he never subscribed to a narrow reading of Genesis. And H. L. Mencken? Well, he was a newspaperman, and his job was to tell good stories and sell newspapers. (For details see Larson 1977; Numbers 1998.)

Finally, Arkansas in the early 1980s. This was a state that was starting to awake from the slumber of the South, but only fitfully. The young and energetic new governor, Bill Clinton, had overreached himself and been thrown out of office after only one term. Chastened and contrite, Clinton was to regain his post some two years later, but the interregnum was filled by a man whose surprise at his election was equalled only by his inadequacy to the job. The bill which insisted on the teaching of creation science in the biology classes of the state's schools was passed, one Friday afternoon, after somewhat less than thirty minutes' debate, and signed into law by the governor, who seems not to have realized quite what it all meant. The real forces of power in the state, represented by the Junior Chamber of Commerce, were absolutely appalled. They had been bending Heaven and Earth to persuade industry to invest in Arkansas and to build factories – usually factories making or assembling high-technology products like computer chips. In their quest to entice some bright young electronics engineer, graduate of MIT, to leave the Boston area and resettle with his wife and family in Little Rock, the last thing they needed was some law ensuring that his children would be taught creationism in science class. He would pick up his family and talents and move farther west, to a state like Arizona, whose evangelical Christians subscribed to a decent and profitable hypocrisy, carefully separating faith and business.

If the Arkansas authorities had mixed feelings about their law, the same was no less true of the ACLU in attacking it: or if their feelings were not mixed they were certainly not pure. The law did violate the U.S. constitutional separation of church and state. There is no question about that. But the enthusiasm of the ACLU for attacking the law far exceeded the legal facts. The association had just come off a wrenching encounter when – entirely properly and honourably, I stress – it had felt obliged to defend the right of American Nazis to march through Skokie, a predominantly Jewish suburb of Chicago. The ACLU had won, and the Nazis had marched. But, although as a norm Jews are some of the strongest supporters of the ACLU – they know only too well what happens when a constitution is violated and laws are perverted – here, in every sense of the word, was something too close to home. Contributions to the ACLU had plummeted and finances were just plain awful. Hence, the organization was desperate for a high-profile case, where it could clearly be seen to be doing good, on behalf of a cause and in a fashion with which its traditional supporters could sympathize. A law supposedly promoting the interests of rabid conservative Christians, in a perceived backward southern state, was manna from heaven, to use a phrase (Gilkey 1985; Ruse 1988a).

In recording the background to these celebrated clashes between Darwinian evolution and Christianity, it is not my intention simply to suggest that there are no genuine points of conflict between biological science and the most popular religion of the West. Anyone who has ever done serious historical research knows full well that people's motives are rarely what they appear on the surface and never mere matters of logic and reason. But I do want to sensitize you to – make you cautious about – stirring tales of the opposition between Darwinism and Christianity. In intellectual matters as in the world of commerce: caveat emptor! If nothing else, you should be aware that there have been many enthusiasts for evolution, Darwinian evolution, who have been sincere Christians. They may be wrong, and those who think that there must inevitably and always be warfare between science and religion may be right, but Christian evolutionists have existed right down from the time of Darwin, and they command our respect if not agreement. Moreover, you should not think that these people were apologetic, trying desperately to bring together their science and their religion but pulling back from any attempt at genuine integration. They have been happy and sincere in both Darwinism and Christianity, seeing them as reflections of a united whole.

One of the very first men to speak publically in favour of Darwinism was not just a Christian but an Anglican clergyman. The Reverend Baden Powell, Savilian Professor of Geometry at the University of Oxford (and father of the future founder of the scouting movement), wrote, even as Huxley was debating Wilberforce, of "Mr. Darwin's masterly volume on *The Origin of Species* by the law of 'natural selection,' – which now substantiates on undeniable grounds the very principle so long denounced by the first naturalists, – *the origination of new species by natural causes:* a work which must soon bring about an entire revolution of opinion in favour of the grand principle of the self-evolving powers of nature" (Powell 1860, 139).

In such sentiments as these Baden Powell was joined by others: on one side of the Atlantic by the Reverend Charles Kingsley, professor of modern history at the University of Cambridge, controversialist in debate with John Henry Newman, and author of *The Water Babies* (Kingsley 1895); on the other side of the Atlantic by the chief spokesman for the Darwinian cause in North America, the professor of botany at Harvard University and ardent evangelical Christian, Asa Gray (1860). Nor did people simply react and emote. They thought carefully about the implications of Darwinism for Christian faith. One of the most attractive and incisive of those who embraced both Darwinism and Christianity was the Oxonian Anglo-Catholic theologian Aubrey Moore. Explicitly, he saw science – particularly Darwin's science about organic origins – as something which brings God intimately and constantly into the picture, at all times and in all places. "Darwinism appeared, and, under the guise of a foe, did the work of a friend" (Moore 1890, 73). It shows us that either God is everywhere or He is nowhere. "We must frankly return to the Christian view of direct Divine agency, the immanence of Divine power from end to end, the belief in a God in Whom not only we, but all things have their being, or we must banish him altogether" (74).

Do not think, however, that it was only those people approaching the science/religion interface from the side of religion who were keen to integrate and harmonize. Near the top of anyone's list of the "ten greatest evolutionists since Darwin" will be the English statistician Ronald Fisher,

author of *The Genetical Theory of Natural Selection* (1930), and the
Russian-born American Theodosius Dobzhansky, author of *Genetics and
the Origin of Species* (1937). Both were ardent Christians, Fisher in the
Church of England of his childhood and Dobzhansky in the Russian
Orthodox Church of his childhood (although in later life he tended to a
more universalistic faith). Listen first to Fisher, from a sermon which he
gave at his college at a Sunday evening service:

To the traditionally religious man, the essential novelty introduced by the theory
of the evolution of organic life, is that creation was not all finished a long while
ago, but is still in progress, in the midst of its incredible duration. In the language
of Genesis we are living in the sixth day, probably rather early in the morning, and
the Divine artist has not yet stood back from his work, and declared it to be "very
good." Perhaps that can only be when God's very imperfect image has become
more competent to manage the affairs of the planet of which he is in control.
(Fisher 1947, 1001)

Evolution is not yet finished, and apparently the task set on us by God is
that of seeing that it does not go astray through blind disregard for the
stern laws of nature.

Now hear Dobzhansky likewise blending science and faith, in a letter
to an eminent historian of science, also a practicing Christian:

I see no escape from thinking that God acts not in fits of miraculous inter
ventions, but in all significant and insignificant, spectacular and humdrum
events. . . . In evolution some organisms progressed and improved and stayed
alive, others failed to do so and became extinct. Some adaptations are better than
others – for the organisms having them; they are better for survival rather than for
death. Yes, life is a value and a success, death is valueless and a failure. So, some
evolutionary changes are better than others. Yes, life is trying to hang on and to
produce more life. (Greene and Ruse 1996, 463)

But, you will say to me: Enough of the past! What of people's beliefs
today? Surely it is the case that science and religion have now truly come
apart and that, even if once they may have had a relationship, as we move
into a new millennium Darwinism and Christianity are well and truly
divorced? At the very least, to use a trendy term which one sees about,
they must be incommensurable. At the very most, if you accept Darwinism then you must reject Christianity, and conversely. The two belief
systems are contradictory.

Well, one thing I cannot deny is that there are those who think that this is precisely the case. Today's most splendid polemicist for pure unadulterated Darwinism is the Oxford biologist of social behaviour Richard Dawkins. He is also a man who takes his atheism seriously, so much so that in contrast the great eighteenth-century Scottish philosopher David Hume (once memorably described as "God's greatest gift to the infidel") looks positively wet. Of those who would compromise or treat the opposition with respect, Dawkins sneers that a "cowardly flabbiness of the intellect afflicts otherwise rational people confronted with long-established religions" (Dawkins 1997b, 397). For him, no quarter is asked and none given. He will have nothing to do with accommodation. "A universe with a supernatural presence would be a fundamentally and qualitatively different kind of universe from one without" (399).

You are either with Dawkins or against him. A point of full agreement with those on the other side. Whatever the successes or failures in Arkansas, there is today an active opposition to all naturalistic claims about organic origins. If Thomas Henry Huxley was Darwin's bulldog, Phillip Johnson, legal scholar and Berkeley faculty member, might well be called creationism's terrier. He worries away at the evolutionists, grabbing and nipping at their ankles, regarding snarls and blows as friendly invitations to further combat. Johnson is as forthright as Dawkins: "Makeshift compromises between supernaturalism in religion and naturalism in science may satisfy individuals, but they have little standing in the intellectual world because they are recognized as a forced accommodation of conflicting lines of thought" (Johnson 1995, 212). Johnson is no mediator. He is a lawyer, for whom agreement means abject capitulation by the opposition. In his book, any compromise is "makeshift."

Yet not everyone thinks this way. There are many who argue passionately that science and religion, Darwinism and Christianity, can coexist harmoniously. Stephen Jay Gould, a paleontologist who is as famous for his popular writings as is Dawkins, is one who sees no conflict between evolution and religion. Raised a secular Jew, he describes himself as an agnostic, but he has always struck me as being closer to God than many conventional believers. At least since the time when he appeared alongside me as an expert witness for the ACLU in Arkansas, Gould has argued repeatedly and vehemently that science and religion do not and (properly understood) cannot clash. "The lack of conflict between science and religion arises from a lack of overlap between their respective domains of professional expertise – science in the empirical constitution of the universe, and religion in the search for proper ethical values and the spiritual meaning of our lives" (Gould 1997b, 18). This same sense of unity and harmony is likewise felt by many who come to science from the side of religion. Keith Ward, Regius Professor of Divinity at Oxford, not only speaks of natural selection as a "simple and extremely fruitful theory," but goes on to say that there is "every reason to think that a scientific evolutionary account and a religious belief in a guiding creative force are not just compatible, but mutually reinforcing" (Ward 1996, 63).

If nothing else, you must agree with me that we have a lively debate here, with disputants spread right across the spectrum. I am not going to solve everything in this book, but I do think that these are interesting and important issues which concern intelligent and thoughtful people, and I assume that since you have read this far you agree with me. My experience as a teacher and a scholar – and as an evolutionist, I might say – is that when you have intelligent people disagreeing, you do best to try to go back to roots and to lay out basic things as clearly as you can. Nothing should be ignored, and nothing should be taken for granted. This is particularly important when, as here, you are dealing with subjects which straddle several different fields of intellectual inquiry: biology, theology, philosophy, and probably history, just for a start.

So, to see if we can move forward on the defining question of my title – Can a Darwinian be a Christian?   I shall now start right back with the basics, trying first to lay out what I think are fairly standard and generally acceptable characterizations of what it is to be a Darwinian and what it is to be a Christian. Then I shall go on to see how they compare and contrast, where we find the points of agreement and where we find the points of tension and perhaps outright disagreement. I stress that my aim here is not to establish whether it is reasonable to be a Darwinian or to be a Christian. I have had much to say elsewhere on some of these matters, and I am certainly not alone in this. For now, I put these questions and battles to one side. My task is deliberately limited, but I hope thereby to achieve a depth that (and I speak now of my past self as much as of anyone else) tends to be conspicuously lacking in discussions of this sort.

CHAPTER ONE

Darwinism

It is useful when discussing anything to do with evolution to make a threefold division. We have the *fact* of evolution, the *path* (or *paths*) of evolution, and the *mechanism* or *cause* (*mechanisms* or *causes*) of evolution (Ruse 1984). By the fact of evolution we mean the development by natural causes of all organisms, those today and those yesterday, from other forms probably ultimately much simpler and originally perhaps from nonliving substances. By the paths of evolution, or (as they are known technically) by the "phylogenies," we mean the actual tracks that evolving lines of organisms took through time. By the mechanism or cause of evolutionism we mean the force (or forces) actually driving the evolutionary process.

Evolution as Fact

The Englishman Charles Robert Darwin (1809–1882) is the key figure in the history of evolutionism and the key figure in our story. He was not the first evolutionist, in the sense of being the first to propose and argue for evolution as fact. Back in the eighteenth century, his grandfather Erasmus Darwin had believed in naturally caused organic origins, and later there were others, including the early French evolutionist Jean Baptiste de Lamarck (1809), who argued that organisms change because characters acquired during an organism's lifetime can be inherited by descendants (McNeil 1987; Burkhardt 1977). However, Charles Darwin in his *On the Origin of Species* (1859) was the person who made evolution more than just an idea – more than just a sweeping hypothesis which sounds good but which has little evidential support. Let us start right here with our discussion of evolution as fact. (A review of the Darwin literature can be found in Ruse 1996b.)

Darwin was a fortunate man. He was educated at the University of Cambridge just at the time when leading thinkers of his day, notably the astronomer John F. W. Herschel (1830) and the historian and philosopher of science William Whewell (1837, 1840), were articulating conditions for scientific excellence. Those two men agreed that a crucially excellent form of scientific argumentation involves what Whewell (who rather liked inventing terms and phrases) called a "consilience of inductions." In order to establish a claim or hypothesis, one must appeal to a wide range of evidence: the evidence being explained by the hypothesis and the hypothesis in turn being supported by the evidence. "The cases in which inductions from classes of facts altogether different have thus *jumped together,* belong only to the best established theories which the history of science contains" (Whewell 1840, 2, 230). Both Herschel's and Whewell's favourite example was Newtonian mechanics, where the central claims – particularly that about the force of gravitational attraction – explain the movements of the planets in the heavens (Kepler's laws) as well as the motion of projectiles here on earth (Galileo's laws) and in turn are given support by such explanatory success.

The consilience of inductions is the crucial tool in the historical sciences. A prime example is the revolution in geology which took place nearly forty years ago, when it was established that the continents "drift" around the planet. The theory of "plate tectonics" is used to explain such phenomena as the deep rifts or channels in the middle of oceans and the geographical locations of areas of high volcanic and earthquake activity, and conversely these phenomena support the theory of continental movement (Ruse 1989). But the paradigmatic example of a consilience is the "one long argument" (as Darwin himself called it) of the *Origin.* Explicitly guided by what he had learnt from Herschel and Whewell, Darwin gathered the evidence for the fact of evolution from right across the biological spectrum (Figure 1). Biology was explained and evolution supported (Ruse 1975a,c, 1979a).

Paleontology: There is a roughly progressive fossil sequence up from extinct forms to remains of organisms hardly different from those

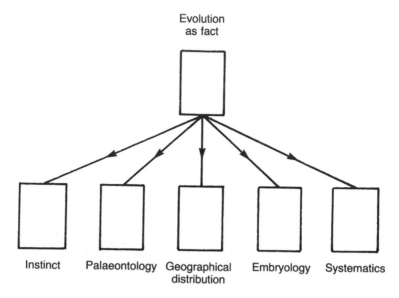

1. The structure of Darwin's argument for the fact of evolution. The fact explains and unifies claims made in the subdisciplines (only some of which are shown), which in turn yield the "circumstantial evidence" for the fact itself. (Ruse, M. *Taking Darwin Seriously,* 1986.)

we see around us today. What better answer than an evolutionary rise from first to last? (Figure 2)

Biogeography: Animals and plants are not scattered randomly around the globe. The birds and the reptiles of the Galapágos Archipelago, for instance, are similar but different from island to island and similar to but different from those of the mainland. What other explanation than that ancestors came from the mainland and diversified once they arrived at the island group? (Figure 3)

Anatomy: Organisms of quite different species have bones and other bodily parts which are isomorphic to each other, despite their completely different functions. The standard example is the forelimb of the mammal: the human arm used for grasping, the horse's leg used for running, the seal's flipper used for swimming, the bat's wing used for flying, the mole's paw used for digging. How else can one explain these similarities (known technically as "homologies") in a natural way other than by evolution? (Figure 4)

Systematics: Why should one find that organisms arrange themselves

TABLE of STRATA and Order of Appearance of Animal Life
upon the Earth.

TERTIARY or CÆNOZOIC			
Tuibary. Shell-Marl. Glacial Drift. Brick Earth. } Bone-Caves.	Pleistocene	MAN by Remains, ... by Weapons.	Birds and Mammals.
Norwich Red Coralline } Crag.	Pliocene		
Faluns. Molasse.	Miocene	Ruminantia. Quadrumana. Proboscidia. Birds, Orders of. Rodentia. Mammals, Orders of.	
Gyps. London } Clays. Plastic }	Eocene	Ungulata. Carnivora.	

SECONDARY or MEZOZOIC			
Maestricht. Upper Chalk. Lower Chalk. Upper Greensand. Lower Greensand.	Cretaceous	Cycloid. } Fishes. Ctenoid. } Mosasaurus. Polyptychodon. Birds, by Bones. Procœlian Crocodilia.	Reptiles.
Weald Clay. Hastings Sand. Purbeck Beds. Kimmeridgian. Oxfordian. Kellovian.	Wealden / U.	Iguanodon. Marsupials, — Chelonia by Bones. Pliosaurus.	
Forest Marble. Bath-Stone. Stonesfield Slate. Great Oolite. Lias. Bone Bed.	M. Oolite / L.	Marsupials. Amphicœlian Crocodilia. Pterosauria. Homocercal Fishes. Cephalopods 2-gilled. Icthyopterygia.	
U. New Red Sandstone. Muschelkalk. Bunter.	Trias	MAMMALIA AVES, by Foot-prints. Sauropterygia. Labyrinthodontia. Crustacea 10-porta.	

PRIMARY or PALEOZOIC			
Marl-Sand. Magnesian Limestone. L. New Red Sandstone.	Permian	Sauria. Chelonia, by Foot-prints. Isopoda.	Fishes.
Coal-Measures. Mountain Limestone. Carboniferous Slate.	Carboniferous	REPTILIA ganoceph. Insecta.	
U. Old Red Sandstone. Caithness Flags. L. Old Red Sandstone. Ludlow.	Devonian	PISCES { ganoid. placo-ganoid. placoid. Heterocercal.	
Wenlock. Caradoc. Llandeilo. Lingula Flags. Cambrian.	Silurian	Echinoderms. Annelids. Bivalves. Trilobites. Pteropoda. Brachiopods. Gastropods. Cephalopods 4-gilled. Fucoids. Zoophytes.	Invertebrates.

2. The animal fossil record, as known at the time of the *Origin*. By this time, all serious scientists (including nonevolutionists) realized that the history of life was not a single progressive rise, but involved much branching. (Owen, R. *Paleontology*, second edition, 1861.)

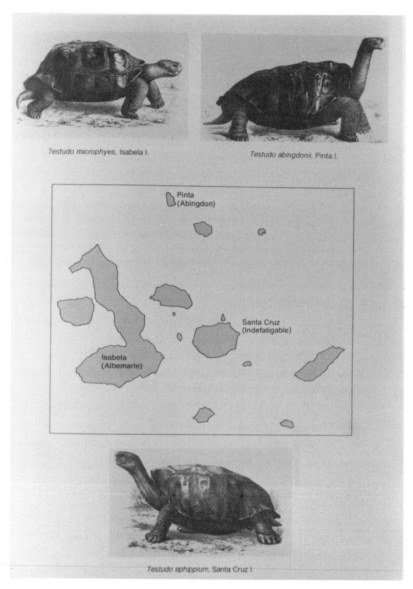

Testudo microphyes, Isabela I.

Testudo abingdonii, Pinta I.

Pinta
(Abingdon)

Santa Cruz
(Indefatigable)

Isabela
(Albemarle)

Testudo ephippium, Santa Cruz I.

3. Three different tortoises from three different islands of the Galápagos. (Dobzhansky et al. *Evolution,* 1976.)

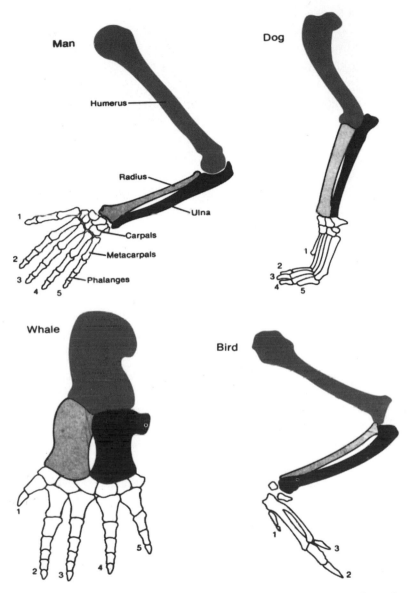

4. Homology between the forelimbs of several vertebrates. Numbers refer to digits. (Dobzhansky et al. *Evolution,* 1976.)

in groups, which can be ordered hierarchically, with many groups (species) at lower levels and more and more inclusive groups as one rises through the ranks (from species to genus, and so on up to kingdoms)? What better answer than that this ordering reflects common ancestry? (Figure 5)

Embryology: Organisms of different species with adults of very different forms have embryos which are identical. The naked eye cannot tell apart the embryos of dog and human. This points to a shared evolutionary origin. (Figure 6)

And so the story goes on, until the case for evolution as fact was put "beyond reasonable doubt." We have, as Darwin said, a great tree of life, "which fills with its dead and broken branches the crust of the earth, and covers the surface with its ever branching and beautiful ramifications" (Darwin 1859, 130) (see Figure 9, p. 184)

Evolution as Path

For all that he used the metaphor of a tree of life, Darwin himself had relatively little to say about the actual path of evolution. There is virtually no discussion of the topic in the *Origin*. He did have thoughts on the subject, but generally these were implicit rather than expressed. Phylogeny was clearly much on his mind in an earlier (around 1850) massive study he did on the barnacles, but at that point he was not about to reveal his evolutionary hand in any direct fashion (Darwin 1851a,b, 1854a,b). Yet although Darwin said little, by the middle of the nineteenth century others were finding and confirming the basic outlines of Earth's history. The fossil record was being uncovered, and it was being realized that it has an interpretable pattern, from earlier marine forms (like trilobites), up through the vertebrate fish, amphibia, reptiles, and then on to birds and mammals (Bowler 1976). (We now date the big explosion of life, the Cambrian period, from about 550 million years ago. This is a mere fraction of earth's history, which is about 4.5 billion years, or even of the history of life, which is about 3.7 billion years.)

Even as the *Origin* was being published and debated, the fossil record was yielding its treasures: petrified forms which pointed to evolution and which showed the specific paths. In the late 1850s, interesting fossil feathers were being discovered in mines in Germany; but it was not until

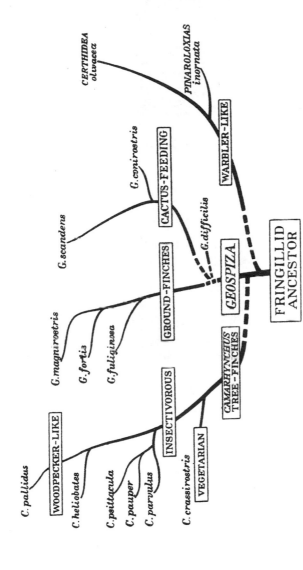

5. Suggested evolutionary tree of Darwin's finches, showing relationships. (Lack, D. *Darwin's Finches*, 1947.)

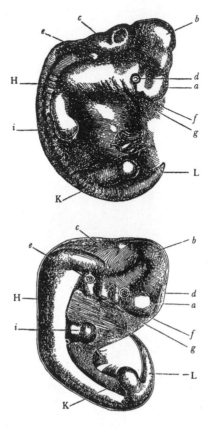

Fig. 1.   Upper figure human embryo, from Ecker. Lower figure that of a dog, from Bischoff.

a. Fore-brain, cerebral hemispheres, etc.
b. Mid-brain, corpora quadrigemina.
c. Hind-brain, cerebellum, medulla oblongata.
d. Eye.
e. Ear.
f. First visceral arch.
g. Second visceral arch.
H. Vertebral columns and muscles in process of development.
i. Anterior }
K. Posterior } extremities.
L. Tail or os coccyx.

6. Comparison of human and canine embryos. Why are they so similar if the human and the dog do not share a common ancestor? (Darwin, C. *The Descent of Man*, 1871.)

the next decade that the whole organism was discovered and it was seen that these feathers belonged not to a normal bird, but to a bridge across the gap between reptiles and birds. Archeopteryx, with its reptilian brain and separate digits and tail and teeth, but also with the apparatus for flight, was just the "missing link" that evolutionism demanded (Feduccia 1996). Yet, exciting as a discovery like this truly was, the greatest glory of phylogenetic recovery and reconstruction was yet to come. In the second decade after the *Origin,* across the Atlantic in the American West, fossil hunters discovered immense charnel houses of the past, bringing to light all of those incredible vertebrates which populated the earth in earlier days: the huge dinosaurs particularly, as well as incredible later mammalian monsters like the gigantic titanotheres, rhinoceros-like brutes with fabulous baroque appendages to their snouts. Simplest, but most wonderful of all, were the equid discoveries, making it possible to trace the modern single-hoofed horse right back to a dog-sized creature, Eohippus, which ran around the prairies on its five toes. This really was phylogenetic revelation with a vengeance (Osborn 1910, 1931; Bowler 1996).

Not that fossils were the only guides to life's history. Other techniques for ferreting out the past involved comparisons between anatomical features, and increasingly the use of embryological analogies and dissimilarities. Even though superficially organisms may be very different, underlying homologies – particularly those in the embryonic forms – might be most revealing about connections and similarities, which could then be interpreted with regard to origins (Russell 1916). One could indeed work in this way, even in the absence of any fossil evidence whatsoever. I should say that, although Darwin used these ideas and methods extensively in his barnacle studies, they were never particularly Darwinian as such. Indeed, in themselves they were not particularly evolutionary ideas, nor have they ever become so. The early nineteenth-century idealistic (or transcendental) German philosophy known as *Naturphilosophie* was ever the chief influence (Gould 1977b; Richards 1987). These thinkers sought connections and harmonies throughout nature, seeing anatomical similarities as evidence of the World Spirit. If one wants to put this in evolutionary terms, then so be it. Conceptually, however, it is not necessary.

Evolution as Cause

Returning from the continent to Britain, we come finally to evolution as mechanism or cause. Here, Charles Darwin came fully into his own. Although he accepted the inheritance of acquired characteristics – never rejected it, in fact – Darwin realized that it would not support a full evolutionary causal perspective. In 1838, after an intensive search, he hit upon the mechanism which would figure prominently in the *Origin:* "natural selection" or (as it came to be called following a suggestion of his fellow English evolutionist Herbert Spencer) the "survival of the fittest." Working by analogy from the ways in which animal and plant breeders change and improve forms – that is, by selecting and breeding only from the best and most desirable stock – Darwin argued that there is a similar process occurring in nature (Ruse 1979a). But how is this brought about, if not by the conscious intention and design of a thinking being? Simply by the force of population, as had been highlighted some years earlier by the political thinker the Reverend Thomas Robert Malthus (1826). Population growth is bound to outstrip available resources, and there will be a consequent "struggle for existence."

A struggle for existence inevitably follows from the high rate at which all organic beings tend to increase. Every being, which during its natural lifetime produces several eggs or seeds, must suffer destruction during some period of its life, and during some season or occasional year, otherwise, on the principle of geometrical increase, its numbers would quickly become so inordinately great that no country could support the product. Hence, as more individuals are produced than can possibly survive, there must in every case be a struggle for existence, either one individual with another of the same species, or with the individuals of distinct species, or with the physical conditions of life. It is the doctrine of Malthus applied with manifold force to the whole animal and vegetable kingdoms; for in this case there can be no artificial increase of food, and no prudential restraint from marriage. (Darwin 1859, 63)

Then from this, together with the naturally occurring variation which his study of domestic organisms had convinced Darwin was a universal feature of life, a form of selection analogous to that practiced by breeders could be obtained.

Let it be borne in mind in what an endless number of strange peculiarities our domestic productions, and, in a lesser degree, those under nature, vary; and how strong the hereditary tendency is. Under domestication, it may be truly said that the whole organisation becomes in some degree plastic. Let it be borne in mind how infinitely complex and close-fitting are the mutual relations of all organic beings to each other and to their physical conditions of life. Can it, then, be thought improbable, seeing that variations useful to man have undoubtedly oc-curred, that other variations useful in some way to each being in the great and complex battle of life, should sometimes occur in the course of thousands of generations? If such do occur, can we doubt (remembering that many more individuals are born than can possibly survive) that individuals having any advan-tage, however slight, over others, would have the best chance of surviving and of procreating their kind? On the other hand we may feel sure that any variation in the least degree injurious would be rigidly destroyed. This preservation of favour-able variations and the rejection of injurious variations, I call Natural Selection. (80–81)

One thing must be emphasized now and throughout. Darwin did not look upon natural selection simply as a causal force for evolution. It was a force for evolution of a particular kind. Darwin believed that the most fundamental aspect of organisms is their functional complexity. They are not simply thrown together randomly. They work, they function, they have features – adaptations – which enable them to survive and repro-duce. It is a key premise in the *Origin* that natural selection is precisely the force which brings about such adaptedness and functioning. Darwin did not have much idea about the nature of the new variations which were the building blocks on which selection would fashion adaptation, but he was adamant that they could not be in any sense guided or directed towards the needs of their possessors. In this sense they are "random." It is selection which does all of the work. One consequence that Darwin drew from this is that although variations come in all sizes and shapes, the ones important for selection have to be small: large variations ("hopeful monsters") would take organisms out of adaptive focus with their surroundings and lead to death before selection could make something of them. This means that, for Darwin, evolution has to be gradual and smooth. There is no place for jerks or jumps. Nothing that Darwin said denied that evolution may be faster or slower at some times, but on a generation-by-generation basis the change has to be gradual.

Part of Darwin's argument for natural selection was direct. There is a struggle for existence. There are variations in every population studied carefully by biologists. (He believed this from the beginning, in the 1830s, but Darwin's work on barnacles convinced him that he had not argued improperly.) Natural selection is a deductive consequence. Part of Darwin's argument for natural selection was analogical. Artificial selection is incredibly powerful and effective. One should therefore expect natural selection to exist, and to be no less powerful and effective – to be far more powerful and effective, in fact, since it is operating flat out all of the time. And part of Darwin's argument for natural selection was bound up with his consilience argument for the very fact of evolution. He did not tease apart the various strands of his argument, and so we find that he is always running together the case for evolution as such and the case for the underlying mechanism (Ruse 1973, 1975a).

What about the most interesting organism of them all? What about *Homo sapiens?* Darwin was always convinced that humankind is part of the evolutionary scenario. Unlike so many of his Victorian contemporaries, for him this was never a matter of tension. In the *Origin,* however, Darwin said virtually nothing, thinking it better to get the general account of evolution on the table, as it were. He dropped only the provocative comment: "Light will be thrown on the origin of man and his history." It was later that Darwin moved to develop his thinking with respect to our species, especially in his *Descent of Man* (1871) and then in what was essentially a supplementary work, *The Expression of Emotion in Man and the Animals* (1872).

In connection with human evolution, Darwin did go on to develop his thinking, using a modification of his main mechanism of change. This was a second form of selection, which goes back right to the earliest thinking but which had hitherto been largely neglected. "Sexual selection" is a mechanism supposedly brought on less by the brute struggle for existence and more by within-group competition for mates. The main thing which this invocation of sexual selection does serve to emphasize is the extent to which Darwin's thinking about natural selection was governed by an individual (rather than group) perspective (Ruse 1980). He saw the struggle as being between one individual and another rather than between one group (like a species) and another. This means that adaptations never serve the group, except indirectly. They exist always for the benefit of the individual.

After Darwin

People accepted evolution as fact rapidly and readily (Ellegård 1958; Bowler 1984). They even accepted that evolution applies to our own species. The similarities between humans and the higher apes just seemed too much to ignore or explain away. Finding the paths of evolution, phylogeny tracing, became almost a national pastime, although I should say that this enthusiasm for phylogeny tracing did not last. Well before the century's end, people realized that either there was too little information – or too much! The fossil record was rarely sufficiently detailed to support definitive phylogenies, and when one added in homologies and embryology, contradictions had a nasty way of appearing (Ruse 1996a). Notorious was the so called biogenetic law, supposing that individual development (ontogeny) is a cameo version (recapitulates) historical development (phylogeny). Unfortunately, for all that this seemed to give a powerful way into the history of life on this globe, there was exception after exception, making a mockery of its pretensions. Ambitious young biologists abandoned the quest for evolution's past, seeking scientific glory elsewhere.

What about natural selection? No one denied that it exists or that it works, or even that it has had a role in human evolution. But most of Darwin's contemporaries refused to allow that it can have the effects supposed in the *Origin* and the *Descent* (Bowler 1983). Those who agreed with Darwin that adaptation is a significant feature of organic nature (like his supporter in North America, the Harvard botanist Asa Gray) opted for souped-up "Lamarckism" (inheritance of acquired characteristics) or for special designlike raw variations (Gray 1860). Those who denied that adaptation is all that important (like Darwin's great supporter in Britain, Thomas Henry Huxley) opted for large variations ("saltations") or similar mechanisms (Huxley 1893a).

The troubles here were not simple prejudice, nor were Darwin's opponents necessarily antiscientific. A major problem was that of heredity. For all that Darwin believed that variation is abundant and that much of it is small and none of it directed, he really had very little idea about the causes of such variation or of its transmission from one generation to the next. And like most everyone else of his age, he was crippled by the assumption that in each generation the tendency is to blend together the features inherited from parents: thus, however good a variation may have been from the perspective of selection, its effects would very soon be watered down to an ineffectual level. The breakthrough here was not to come until the turn of the century, when the ideas of an obscure Moravian monk, Gregor Mendel, were rediscovered and developed. Then it was seen that new variations do not necessarily blend away, can be carried intact indefinitely through the generations, and so can give new life to natural selection.

The key conceptual steps came at the hands of mathematically talented biologists. In Britain Ronald Fisher (1930) and J. B. S. Haldane (1932) and in America Sewall Wright (1931, 1932) showed that selection and Mendelian genetics (as the new theory of heredity was known) are complementary, giving a fuller theory and more satisfactory causal theory than Darwin had been able to produce. There were differences among these "theoretical population geneticists" – between Fisher and Wright, in particular, with the former preferring to think of selection as working in large populations and the latter in terms of small groups and beneficial effects spreading to the whole – but the similarities were more important (Provine 1971). Now one could show that good variations would not be swamped right out in a generation or two, and one could and did show how selection could get to work leading to significant permanent change. And from this theory, a number of empirically minded biologists – notably E. B. Ford in Britain and Theodosius Dobzhansky in the United States – set about providing the factual evidence to back up the calculations (Ruse 1996a).

Thus around 1940 was born the modern theory stemming from the blend of Darwinian selection and Mendelian genetics. Classics of the so-called synthetic theory of evolution or neo-Darwinism include (in Britain) *Evolution: the Modern Synthesis* (1942) by Julian Huxley, the grandson of Thomas Henry Huxley, and (in America) *Genetics and the Origin of Species* (1937) by Dobzhansky, *Systematics and the Origin of Species* (1942) by the ornithologist and systematist Ernst Mayr, *Tempo and Mode in Evolution* (1944) by the paleontologist G. G. Simpson, and *Variation and Evolution in Plants* (1950) by the botanist G. L. Stebbins.

Modern Advances

The coming of the synthetic theory set the stage for sustained evolutionary study, in theory and in the empirical domain (Ruse 1982). More has been done on the evolutionary front in the past half-century than ever before. A major stimulus has been molecular biology, a subject marked by the discovery in 1953 of the double helical shape of the DNA molecule by James Watson and Francis Crick. This led to many new ideas and techniques on the evolutionary front (Lewontin 1974; Freeman and Herron 1998). Another major area of advance centres on the study of social behaviour ("sociobiology"). This will be the subject of discussion in a later chapter. Other points of advance (and controversy) are in the realm of paleontology, as new models are devised and as the earth discloses its secrets. And yet other areas of forward movement take us back to traditional problems, including a topic that intrigued Darwin from the first and which made its way into the *Origin:* the evolution of animals on oceanic islands, specifically the evolution of the birds on the Galápagos Archipelago in the Pacific. Why do we find that on a small group of islands, most within sight of each other, there is such avian diversity? Why do the finches (now known as "Darwin's finches") have so many (thirteen) different species, and why do they show such different adaptations for feeding? Some are cactus and seed eaters, some are insectivorous, some eat fruit, and there are even two species that pick up twigs in theirs beaks and poke around in bark for insects!

Researchers have done extensive studies on the medium ground finch (*Geospiza fortis*) on the islet of Daphne Major (Grant 1986; Grant and Grant 1989). They have counted all of the birds on the islet (around a thousand or so), traced lineages so that they could tell who was passing on which characteristics to whom, monitored the effects of drought and abundance of rain, checked on the availability of foodstuffs (the finch is a seed eater), and seen which finches are eating which resources, which ones are getting plenty and which little. Their results show unambiguously the working and effects of natural selection. For instance, in 1977 there was a major drought on the islet, killing over 80 percent of the finch population. Those that survived were those with stronger beaks, because they could crack the available large seeds. Expectedly, there was a shift in beak form towards larger size, something correlated with the selective forces and the fact that finches with strong beaks are precisely those that have offspring with strong beaks. Showing, however, how this shift could be reversed, in another year with heavy rain and abundant feeding material there occurred a selection-driven move to smaller body size – correlated with less robust beaks. Now selection pushes things one way, now another.

This is a well-known study, one with sentimental appeal to students of Darwinism. It is but the tip of an iceberg, as many, many similar studies are performed today. Increasingly the tendency is to move to small, smaller, and even microorganisms, for these breed rapidly and thus lend themselves to selection experiments. It cannot be denied that, however one may feel about Darwinian evolutionary theory, it is indeed a thriving field of work.

Defining Darwinism

So what is meant by "Darwinism"? Who today is a "Darwinian"? At the most basic level obviously, one is going to accept evolution as fact. That is essential. Organisms living and dead came by a long, slow, natural (law-bound) process from primitive forms, which lived long ago. At the level of path, the Darwinian expects to see life as a tree of some form: whether the tree be a cedar or poplar or oak is beside the point. Branching is important in the divergence of forms, although most agree that a stem could change quite drastically without branching. More than this, the Darwinian as such would not specify. Individual Darwinians might differ dramatically over such things as the evolution of the birds. Did they come from dinosaurs or from other sorts of reptiles? But this would be an in-house quarrel, and not at all threatening to ultimate commitments and allegiances.

What about mechanisms or causes? A Darwinian has to take natural selection seriously. Indeed, a Darwinian has to regard natural selection as the most important evolutionary mechanism that there is. Are there working evolutionists who would deny this? Probably not many, although there is disagreement on how important "most important" really is. No one from Darwin on has ever wanted to claim that natural selection is all-important. For a start, there is sexual selection. But, even after we have rolled all kinds of selection into one, no one has wanted to claim that it alone fuels organic change. There are, as there have always been, alternatives and supplements to natural selection. What is the case is that there has been a change since Darwin in these alternatives and supplements. Once, Lamarckism – the inheritance of acquired characteristics – was considered a very significant secondary mechanism. Today, this process is simply discredited, absolutely and completely. Mendelian/molecular genetics negates it entirely.

So what are today's viable potential alternatives? They include: *dif-ferential growth* (allometry), where one part of an organism grows more rapidly than other parts and perhaps even moves into a nonadaptive phase (it is suggested that the antlers of the Irish elk might be such a case, having been selected in fast-growing juveniles even though by the time the animal is fully grown they are maladaptive); *constraints* on develop-ment, where selection simply has to work with the materials at hand and cannot transgress certain boundaries; *pleiotropy,* which means that one gene (or gene complex) has multiple effects, and although some of these are highly adaptive, others are much less so; and *genetic drift,* a notion which was introduced by Sewall Wright, suggesting that in small popula-tions the effects of random sampling might outweigh those of selection.

These possible alternatives and supplements gives us a range of Darwinisms and Darwinians. At the one end of the spectrum, we have the "ultra-Darwinians." The turn-of-the-century evolutionist Raphael Weldon (1898) was a charter member, and so was Fisher (1930) a little later. Today Richard Dawkins (1986) fits the bill, although he is far from alone. For them, selection is what you look for first and primarily. Even nonselective causes and effects are generally thought to be selection-connected: something may have no selective value now, but it was selection-based in the past. The four-limbedness of vertebrates would be a case in point (Maynard Smith 1981). Four rather than six (like the insects) may have no direct adaptive value – may be a constraint in some sense – but the first vertebrates were aquatic, and two limbs front and two limbs back was a crucial part of being able to move up and down in the water (as an airplane today needs something similar, fore and aft, to move up and down in the air). With this kind of Darwinism you expect to find adaptation, now or in the past, directly or indirectly, and you fight tooth and nail until you find it. These Darwinians think that selection applies to humans no less than to the rest of the organic world, although they may well allow that humans require special treatment. Fisher, for instance, thought that upper-class humans are biologically less fit than the lower classes and that remedial action should be taken. Dawkins (1976) has a theory of "memes," analogous to genes, which are supposedly responsible for mental evolution.

Moving across the spectrum, less committed to the all-conquering power of selection, but still firmly Darwinian, would be those who think that selection is most important but that other factors could well be significant. The English evolutionist John Maynard Smith (1981) has confessed to a liking for hopeful monsters. In America particularly, there is a fondness for genetic drift as a supplementary mechanism of some significance. One context in which such drift is often thought very important is that which obtains when a small group of organisms finds itself isolated on an island or similar territory. Because of the variation within populations and because any such small group will therefore necessarily be atypical, it is thought that this will lead to very rapid evolution (Mayr 1954). This hypothesis about a "founder principle" is controversial and not accepted by all, but it is a major part in many people's picture of the evolutionary process. (See Coyne et al. 1997, for criticism.)

Then, as we continue towards the other end of the Darwinian range, we find those who are firmly evolutionist, who think selection very important, but who deplore the excesses (as they see them) of ultra-Darwinians. The Harvard biologists Richard Lewontin and Stephen Jay Gould fit here. They argue that much of evolution is chance or a function of random factors or of constraints on development. They make the well-known claim that many organic features are simply by-products, akin to what they call "spandrels," the spaces left over at the tops of pillars used in medieval churches to keep ceilings from falling (Gould and Lewontin 1979). In a way, often their criticisms seem more semantic than anything else – ultra-Darwinians might well agree that spandrels had no original function, even if now they play a role – but there is a significant psychological difference separating these evolutionists from the ultras. This is coupled with reluctance to see humankind as simply one with the rest of the organic world (Gould 1981; Levins and Lewontin 1985). These critics simply look at organisms differently than do the ultra-Darwinians: they have to be convinced that selection is at work, rather than assuming that it must be at work.

Since many of these people, Lewontin especially, work from a Marxist perspective, there is a reluctance to tie humans – particularly consciousness – too firmly to selection and adaptation. Selection was vitally important in our past; no evolutionist today is going to deny that. But it would be thought, especially as we approach modern humankind, that consciousness and its products – language and culture in particular – gain an autonomy of their own. They become independent of selection, and one should not look too strongly for adaptation as the defining and controlling factor in thought and its consequent action.

Along with the Marxism, one often finds a fondness or outright enthusiasm for *Naturphilosophie*. Gould especially sees isomorphisms –
homologies – as being significant, at least as significant as adaptation
(Gould 1977b, 1982, 1997a). For this reason there is a liking of *Baupläne*,
that is, blueprints or ground plans which structure the members of a
group. With this, especially considering evolution over long periods of
time (Gould is a paleontologist), comes a willingness to see evolution as
proceeding more in fits and starts than one might expect (anticipate)
given orthodox Darwinism. There is not the need or pressure to stay in
tight or well-defined adaptive focus. Well known (notorious even) is
Gould's theory of "punctuated equilibria" (formulated with fellow paleontologist Niles Eldredge). the claim that evolution consists of long
periods of little or no change (stasis) broken (punctuated) by rapid movement from one form to another (Eldredge and Gould 1972; Gould and
Eldredge 1977).

By now one might question whether people like Gould and Lewontin
should be called Darwinians at all. The answer is surely that it is a matter
of context, as their own usage suggests. Compared to creationists, Gould
and Lewontin are not just evolutionists but Darwinians, and proud to call
themselves such (Gould 1977a). Selection is a very important component
in their evolutionary thinking. But when arguing with fellow evolutionists, they do not feel particularly Darwinian, and feel they have good
reasons for keeping a zone of separation. Nonetheless, one should note
that even further along the spectrum there are other evolutionists who
are more extreme than Gould and Lewontin and their sympathizers.
There are some who argue that selection is fine as far as it goes but that
there is much in the living world that lies beyond the scope of this
mechanism. In particular, it is claimed that at the molecular level, below
the grasp of selection, one has a great deal of random, directionless
change. Evolution in this respect is "non-Darwinian."

How significant all this might be is a matter much contested today.
There are findings suggesting that selection operates even down to the
level of the basic particles of life. Moreover, it should be noted that even
the greatest enthusiasts for "molecular drift" often concede – even firmly
insist – that at the larger, grosser level of being, selection is the truly
significant causal factor in evolution (Kimura 1983). Here they stand
apart from other critics of ubiquitous selection who think that there are
physical forces more significant than hitherto recognized. Selection per-

haps is a kind of surface effect, significant in its way, it is true, but at another level nature itself possibly has certain tendencies towards complexity or organization of certain kinds, irrespective of any biological factors like natural selection. For them, change is going to emerge naturally, irrespective of its subsequent success in the reproductive marketplace (Kauffman 1993; Depew and Weber 1994).

Enough! The aim of this chapter is to provide a conceptual geography, not a listing of every dissident who thought that he or she could better one of the greatest achievements of Western science. I must stress for the record that the number of dissidents among working evolutionists is minute, and their visibility (such as it is) greatly exaggerated by the fact that dissent does garner more attention than assent and staying with the norm. Rarely do their speculations leave the computer screen and find their way into the laboratory or field station. Working evolutionists, looking at real organisms, stay within the Darwinian fold: natural origins, a branching tree, selection and adaptation.

CHAPTER TWO

Christianity

Christianity is a religion, something which claims to explain ultimate reality and the place and role of humans therein. It is something with a social structure, and it lays moral prescriptions upon its members, with offers of reward and (particularly in the past) threats of punishment. Again, I will start the discussion historically, and again I note that not everyone means the same thing by "Christianity." Because my intent is to compare Christianity to Darwinism, a product of Western thought, I will not apologize for an exclusive emphasis on Western Christianity. (In what follows, the backbone of the discussion was provided by Jaroslav Pelikan, *The Christian Tradition* [1971–89]. Also useful have been Metzger and Coogan 1993; Hodgson and King 1994; Gunton 1997; and McGrath 1997.)

Jesus Christ

For the Christian, the essential and central event is the existence, the life and death, of Jesus of Nazareth, a Jewish preacher and prophet and healer who lived two thousand years ago, in Palestine. His behaviour and sermons so upset the authorities, especially the conservative Jewish religious authorities, that he was seized and crucified by the Romans, the occupying power. It is the Christian's claim that Jesus was no ordinary person but that he was the son of God – the Messiah foretold by the Jews, the Christ – and that after death he rose again revealing himself to his disciples before ascending into heaven. The resurrection of Jesus was the ultimate miracle – in a sense, a break with the usual order of nature –

although one of a series that he had performed, from the rather mundane and domestic (turning water into wine at a marriage festival when supplies ran dry) to the truly significant and powerful (raising his friend Lazarus, as well as the daughter of a synagogue ruler, from the dead).

Christianity, like Judaism before it and Islam after, is strictly monotheistic. Although God became human in the person of Jesus, it is an essential part of the faith that Jesus is not separate from God but in some sense is God. Indeed, Christianity is Trinitarian, believing that God has left His spirit – the Holy Ghost – to be with us now that Jesus is gone. Just how we are to regard the exact relationship between Jesus and God is much debated: it is a point of division between the Western branches of Christianity and the Eastern branches, which split about a millennium after Christ (1054). What is not acceptable (in traditional Christianity) is to argue either that Jesus was simply God and not human (and hence did not truly suffer death on the Cross) or that he was purely man and hence, however good and however much a guide, fundamentally mistaken in his claims to divinity.

Why did Jesus die, or rather, what was the significance of his death and subsequent resurrection? Here we go back to Judaism, out of which Christianity arose. God is seen as creator of Heaven and Earth, ex nihilo: that is, out of nothing. It was not that God came along and shaped already-existing matter. This belief, that matter is in some sense separate from God, and that here lies the cause of evil and wrong, was that of a sect of the early Christian era called the Manichees, who are thus associated with one of the oldest of Christian heresies. For the Christian, God is creator but not just creator: He is an all-powerful creator who acts out of pure love. The early Jewish vision of God – Yahweh – was of a being also capable of showing other emotions like jealousy and anger.

We humans, who are in some sense made "in God's image," are a special and privileged part of God's creation. We are not just animals. Regretfully, however, we have fallen into a state of sin, thanks to the temptation and fall of the original humans, Adam and Eve. It is an essential part of traditional Christian theology, especially as articulated by Saint Augustine of Hippo (354–430), the greatest of all the early theologians (the "Church fathers"), that all people carry innately the mark of Adam's fall. We are all in a state of "original sin." However, Jesus through his suffering – God letting Himself be put to death – washes away our sin and makes possible our future salvation. He is the sacrificial lamb. Jesus himself, although human, was not tainted by original sin. Such sin is transferred down through the generations by way of the sexual act. The mother of God – Mary – conceived as a virgin.

Christian Beliefs

What is the nature of salvation and the future life? I suspect more Christians than not have thought it was something like life here on earth, only better. I like to think of it as a new Mozart opera every night, with lots of fish and chips at the intermission, and no student papers waiting to be marked when I get home. If the food is wrapped in salacious Sunday newspapers that my parents would not have allowed in the house, then so much the better. But really nothing absolute is said on this matter by Christianity other than that it will be an ecstasy and closeness with God.

Jesus' death opens the way to eternal life, but how is it to be achieved? The accounts of Jesus that we have – the four Gospels in the New Testament, together with later reports from his followers – specify strongly that certain moral demands are made on us. As all know, the main Christian moral commandment is to love one's neighbour as oneself. Beyond this, Jesus did not present or leave a polished philosophical system. He spoke rather in stories or parables, such as that of the Good Samaritan: a stranger and outsider who bound up and cared for a wounded man, after his own people had "passed by on the other side" (Luke 10:31). Indeed, much of Jesus' teaching is directed more at the personal level. To the chagrin of some of his followers, he did not come to foment revolution, especially not against the Romans: "Render therefore to Caesar the things that are Caesar's, and to God the things that are God's" (Matt. 22:21).

Given the moral prescriptions, it might be tempting to think one can buy one's way to salvation, into the Kingdom of Heaven, through good acts. To think this is to accept another heresy, the Pelagian heresy. You cannot bribe God. What He gives, He gives unconditionally from the goodness of His own heart. There is nothing that we sinners can do to merit or justify God's goodness or bounty. We are entirely dependant on His Grace. This is a point made strongly by Saint Paul and stressed by Saint Augustine. At the end of the Middle Ages, at the beginning of the sixteenth century, there was a great division in the Western Christian Church: the Reformation. Although this divide obviously had many causes, social and cultural as much as intellectual, the Augustinian stress on "justification by grace" was the key theological point of dispute.

In the opinion of its critics – a claim hotly disputed then and now – the Catholic Church had fallen increasingly into Pelagianism. Supposedly, it saw the way to Heaven as lying in good works and in such acts as the buying of "indulgences," which were supposed to shorten a post-life, pre-heaven state of suspension – purgatory – that many thought was human fate. A thousand years after Augustine, the great reformers, notably the German Martin Luther (1483–1546) and the Frenchman (although Swiss-based) John Calvin (1509–64), reaffirmed God's majesty and total power. God and God alone gives out, and this is beyond our control. Good action may well be important and it is a mark of the saved – God is hardly indifferent to the goodness of Mother Teresa or to the evil of Adolf Hitler – but it is not a payment given for a benefit received. The giving on the one side, God's side, is infinitely greater than anything which can be given on the other side, our side.

Since God is omniscient – that is, all-knowing – and since it is God's goodness which is decisive regarding who is to be saved and who not, saints and sinners are seemingly predestined to their ultimate fates. Yet at the same time, Christianity is strongly committed to belief in human freedom: we are, after all, made in God's image, and nowhere more so than in our capacity for moral action. The resolution of this paradox will occupy us later. It must be solved, for the notion of freedom plays an important role in the Christian explanation of one of the most trouble-some barriers to belief. If God is all-powerful and all-good and is indeed creator of Heaven and Earth, how then does one account for evil? It is unthinkable that God is responsible for evil – to believe this is to deny one of His essential attributes – and to suppose that He cannot prevent it is to deny another of His essential attributes. Some way out of this conundrum must be sought.

Authority

Christianity is more than just a belief system. It is an organized institution with functionaries: priests or pastors or clergy, divided hierarchically with leaders and lesser lights below. Augustine (again!) stressed, against the so-called Donatists, that the church is more than just its fallible sinning members, and that it remains valid even though the individuals within it fail its ideals and norms and demands. Traditional Christianity takes its cue from a saying of Jesus to his disciple Peter, that his faith would be the rock on which the church would be founded. As Christianity moved its locus from Palestine to the centre of the then-known world, Rome, it became the custom to regard the head of the Roman branch of Christianity – the bishop of Rome – as the head of the church, the Pope. From an early stage it was believed that the Holy Spirit does work through the Pope and that hence in a sense church doctrine (dogma) is an evolving phenomenon: an ongoing revelation from on high. It was believed that the Pope and indeed all of the "ordained" clergy were therefore people with a special status, one conferred by God. In line particularly with comments made by Saint Paul, the great evangelist who started by persecuting Christians and then became the most ardent voice for the new religion, sexual activity was regarded with some suspicion. Although it was never condemned as inherently bad (this was a Manichean position), sexual activity was regarded as a sign of weakness. The clergy were expected to remain celibate.

Much of this ecclesiastical systematizing was challenged by the reformers. It has been said that the Reformation was a clash between those (defenders) who were determined to hold to the Augustinian line against Donatism and those (critics) who were determined to hold to the Augustinian line on justification by faith. There is some truth in this. The Church of England, which broke from Rome for political rather than religious reasons – Henry and Elizabeth were determined to be their own Popes – kept the structure of the church and has ever had a weakness for the Pelagian heresy. It is just not sporting of God to pick out winners and losers irrespective of merit: English gentlemen have always had a strong sense of fair play. But all Protestants (as the reformers became known) – and this included the Anglicans – denied that the clergy have some special holy status and denied that the Pope has authority to announce new dogma, piecemeal.

What then is the source of authority if the church and its officers are downgraded, as it were? Some of the more radical elements of Protestantism – not Lutheran or Calvinist but followers of the Swiss-born Huldrich Zwingli (1484–1531), often referred to as Anabaptists – argued that the only ultimate authority is God speaking to each individual. The Quakers (the British sect closest to Anabaptism) spoke of the "inner light" – "that of God in every man" – which guides each and every one of us. But more conservative or orthodox Protestants found the essential foundation to their faith to be the written word of God, the Holy Bible. One finds God by faith alone (*sola fide*), and one is guided to Him by scripture alone (*sola scriptura*).

Catholics, of course, have and acknowledge the Bible, taking it to be true absolutely and completely. Although it should be noted that the Bible was not something given entire at one point. Indeed, the "canon" was only decided after some considerable debate. There is the Old Testament, the part dealing with the Jews and their history, and the New Testament, the part telling of Jesus (in the four Gospels), of the doings of the early Christians (Acts of the Apostles), and of letters by early church leaders. (Protestants do not accept as canonical all of the books accepted by Catholics; some parts are relegated to the "Apocrypha.") Once the Bible was decided upon it became binding on Christians, as indeed it still is. However, as radicals turned inward, Catholics turned outward. The latter thought (and still think) that the Bible must share place with the church itself – the "dual source" theory of theology (John Paul II 1998, 55). It is significant that the Bible remained in Latin, where only educated persons – the clergy overwhelmingly – could ask directly about contents. It was only in the Counter-Reformation, its beginning signalled by the Council of Trent (1545–63), when the Catholic door was shut firmly after the Protestant horse had bolted, that the Bible was made more available to the believers who remained in the true church.

The Reformation was marked at the beginning by the invention of the printing press, and whether as cause or as effect the Bible became much more readily available as a potential source of authority. It is no surprise that the first action of reformers was to translate the Bible into the vernacular (the common language), and Protestant countries have had a premium on literacy for all. As the status of the priest was downgraded, the individual's relationship to God was made more direct, with the Bible as guide. Expectedly, elements of Catholic faith which have little biblical authority are discarded by Protestants. Particularly noteworthy is the much-diminished role of the Virgin, whose recorded role in the Gospels is slight after the birth of Jesus.

To be fair to Catholicism, one should note that the Counter-Reformation became far more than simply a reaction – a belated reaction – to the Protestant forces. New vigor infused the church, notably the establishment of a powerful new group or religious order, the Jesuits, dedicated to promoting the true religion and answerable only to the Pope. Their heroic and successful missionary work in the New World alone stands as confirmation of their effectiveness. And at the same time there was theological renewal. For instance, there was a fresh understanding and appreciation of the Augustinian notion of grace, and an articulation of it which gave theological depth to the rift with the Protestants. Whereas for Luther grace was something given by God, simply declaring the sinner saved – an act of forgiveness in the face of sinful nature – for the Catholic grace became clearly understood as a process of regeneration, so that one became truly worthy of salvation.

Sacraments

One element of Christian faith which is important is that of the sacrament, a kind of sign or token of God: something requiring an action by believers at certain fixed times or on certain fixed occasions. Catholics and Protestants differ over these, with the former acknowledging more of them than the latter. One Catholic sacrament, for instance, occurs with the creation of a priest, not something of such special religious significance for the Protestant. The central sacrament, acknowledged by both Catholics and Protestants (except for some extreme groups like Quakers, who insist that the whole of life is a sacrament) is that of the Lord's Supper or Eucharist. Jesus' last night with his disciples in Jerusalem was spent celebrating the Passover, that meal that Jews share to mark their deliverance from oppression in Egypt. Jesus exhorted his followers to remember him through a reenactment of the meal, adding somewhat mysteriously that the bread they shared was his body and the wine they shared was his blood.

Precisely what this means has given rise to some two thousand years of discussion and dispute. Traditionally the words were taken absolutely literally, and it was thought that the bread and the wine are changed into the body and blood of Christ. This is in some sense a miracle. It was not until the Middle Ages, however, and the rediscovery of the thought of the Greek philosopher Aristotle, that theologians within the Catholic faith worked out a satisfactory solution to explain the Eucharist. This became known as the miracle of transubstantiation. Crucial to an understanding of this idea is the notion of a substance and the distinction between properties and accidents. Substances are things, and properties are attributes or qualities that the things have essentially, distinguishing a substance of one kind from a substance of another kind. Accidents are qualities that a substance has contingently, that if they were changed would not alter the essence of the substance. Traditionally, in the case of humans, the fact that we are rational is considered essential, a property, whereas the fact that we are bipedal is considered an accident. Humans would not be humans without rationality, whereas if we ran on all fours we could still be human.

It is argued that, in the ceremony (the Mass) that the priest performs to change bread and wine into the body and blood of Christ, the accidents remain unchanged, whereas the properties change and hence the substance changes. Hence, for Catholics it really is now the true Jesus in the Eucharist, even though no amount of physical examination of the material is going to make the slightest difference. Protestants challenge this belief directly, although there are differences among them on the proper alternative. Luther denied transubstantiation and argued rather for consubstantiation. Jesus is still present in the bread and wine in some important sense but not in the way claimed by the Catholics. Luther's analogy (taken from the third century theologian Origen) was of the heating of a piece of iron. The substance (the iron) remains unchanged, but Christ (the heat) is now present. More radical reformers would have none of this and denied that the bread and wine are anything more than symbolic of Christ. We do this in remembrance of him – to concentrate on his suffering and his sacrificial love of us sinners – and there is to be no presence of what is essentially otherwise cannibalism. "The flesh is not present literally and corporally. For if it were, its mass and substance would be perceived, and it would be pressed with the teeth" (Zwingli 1526, 309).

These differences over the nature of the Lord's Supper are indicative of significant differences right across the spectrum. The Protestant's relationship to God is very different from that of the Catholic. For the latter, the church is a major mediating phenomenon: the Pope and the lower clergy, the ongoing creation of doctrine, the signs and symbols and rituals, and much more, are the building blocks of Christianity. For the former, the relationship to God is much more immediate. Ultimately, one is on one's own, facing one's creator directly. It is often said that Protestantism is a return to primitive Christianity; however, this is only true in a sense, for Christianity as a religion did not really exist before the labours of the Church fathers. But it is certainly the case that Protestantism is an attempt to get back to a more direct form of Christianity, where one is responding directly to the crucified Christ – to a person and events which existed at particular points in history – and not to an ongoing evolving institution.

After the Reformation

Yet for all the differences – and the Protestant Reformation was surely a most significant event in the two millennia of Christian history, leading to appalling conflicts intellectual and physical (the Thirty Years War, for example) – one senses looking back over the centuries that this was an in-house quarrel. Both sides were firmly convinced of the authenticity of their God and of the divine nature of His son Jesus Christ. The two sides also shared points of detail and doctrine, for Protestantism grew out of Catholicism: Martin Luther, for instance, was an Augustinian monk and owed much to that greatest figure in church theology, not least to his great emphasis on God's redeeming grace. The differences were about how one can best approach God and work towards eternal salvation, and because the differences were within the family and between siblings, the disputes were particularly bitter.

In the centuries following, Christianity was to face very different challenges, coming from without but also penetrating right inside its walls – where such challenges often found sympathetic listeners. Some of these challenges were social. Whatever its faults, medieval Christianity provided a comprehensive worldview for an essentially stable system, where everyone from monarch to pauper had his or her place. After the Reformation society started to change drastically, often as a cause of or in response to factors triggered directly or indirectly by religious changes. A more educated population, for instance, started to prepare the way for a more industrialized society: a phenomenon which particularly caught fire in Britain in the eighteenth century and then spread to the rest of Europe and to the New World. But in a changed world – one which was driven by capitalism, which led to urbanization, to the factory system, to rapid and ongoing changes in social structure – traditional Christianity, even Protestantism, seemed increasingly outdated and irrelevant. Too often it was associated with the forces of repression and of the conservative status quo. It was not speaking to the needs of the people, particularly the common people and the intellectuals, which latter had their intelligence and training but often little stake in the prosperity of the society or country.

There were also intellectual factors leading to a crisis for conventional Christianity. The Reformation occurred at the time of the scientific revolution. Copernican astronomy, which put the sun at the centre of the universe rather than the earth, was a shocking suggestion, violating the smoothly integrated picture that had been built up by the medieval acceptance of Aristotle's belief in a circular universe, with the earth at the centre and perfect spheres revolving around it. Saint Thomas Aquinas (1225–1274), the greatest of the medieval theologians, whose philosophical synthesis of Aristotelianism and Christianity known as Thomism still officially defines Catholic thinking, had seen a perfect mesh between Greek physics and the biblical story of creation which locates the earth and its denizens – humans particularly – at the centre of God's concern.

Copernicus dislodged the earth from this central place, and although he himself did not advocate an infinite universe, thanks to the absence of perceived stellar parallax the universe was now seen to be thousands of times larger than had been believed hitherto (Kuhn 1957). Psychologically, the earth was much diminished, and the place and status of humans seemed far less certain. Why should He care about a speck of dust somewhere in the vastness of space? When later one added to this the extension of time that was being demanded by new discoveries and speculations in the earth sciences, the God of Abraham and Moses and Jesus seemed a lot less familiar.

The Enlightenment

But science was far from the only, or even the most important, factor in the intellectual attack on Christianity. In line with the way in which he saw the Bible and Aristotle to be in harmony on the nature of the universe, Aquinas had codified our understanding of God into two separate but supportive categories: first, there is revealed religion or theology, that is, the knowledge of God that we get through faith, something which is open to anyone; then, second, there is natural theology or religion, that is, the knowledge of God which we get through reason, something which is open only to the educated and trained. It is always emphasized that revelation is superior to reason, but it is part of Catholic doctrine that natural theology is important and valid. Protestants tended to put less emphasis on reason – Luther was rather rude on the subject – but still natural theology played a crucial role in Protestant thought. However, although revealed and natural religion found fervent supporters, by the eighteenth century – that time and movement known now as the Enlightenment – all forms of and routes to religion were under attack, an attack which continued into the nineteenth century.

Revealed religion, especially the Protestant version, puts a heavy emphasis on the Bible. Unfortunately, as historical scholarship – particularly German historical scholarship – developed, with increased emphasis on archival and similar sources, with willingness to draw anthropological analogies, with greater insights into human psychology, with hermeneutical readings and comparisons of texts, it became increasingly obvious that there are major problems with simple literal readings of the biblical texts. There are contradictions, omissions, additions, changes of style, multiple authorships, and much more. The Bible may be the word of God, but it is certainly not the simple, straightforward word of God. It is a very human collection of documents and must at some level be read in this way. As you might expect, it was not long before this so-called higher criticism was being used for a wholesale critique of revealed Christianity, even to the point of arguing that this shows Christian beliefs to be untenable or fictitious. First there was Ludwig Feuerbach (1841) arguing that religion was all a projection of human nature, part of an urge to find ultimate meaning, but with no objective reality. Then there was Karl Marx (1845) complaining that Feuerbach had not gone far enough, that he had failed to see that religion is part of the social system used to keep the masses in their place. "Feuerbach . . . fails to see that 'religious feeling' is itself a social product" (555.35).

Natural religion or theology fared little better. Initially, after the Reformation, it seemed to be going from strength to strength. First there was Descartes refurbishing the ontological argument, the proof of God's existence originating with Saint Anselm which derives His existence from His essence: God is that than which none greater can be conceived and hence exists necessarily. Then the teleological argument – deriving God from the intricate nature of His works, an inference going back to Plato and much championed by Saint Thomas Aquinas – seemed to fit smoothly with the new findings in science. All of the design which investigation was uncovering seemed to speak directly to His existence and magnificence.

All forms of such argumentation fell under the critical scrutiny of the philosophers. It was Immanuel Kant (1929) who spoke most strongly against the ontological argument, claiming that its flaw is that it treats existence as if it were a predicate like any other – colour, for instance. But once we see (truly) that it is not, and that it functions in a different way – to say that God exists is very different from saying that God is good – then the argument falls to the ground. David Hume (1779) likewise savaged the teleological argument, showing that far from proving anything like the Christian God, it proves as readily that the world is more like a cabbage than an object of conscious design, or alternatively that there was a squad of workers producing not just one world but a whole series of worlds, of which ours is perhaps but one.

Nor is the problem of evil simply something to be brushed lightly under the carpet, free will or no free will. In similar fashion, miracles fell under Hume's critical scrutiny, and it was concluded that it is more reasonable to think of such reports as the results of ignorance and wishful thinking than as narratives of genuine interventions by God in the normal course of nature. As it happens, Hume himself probably did not deny absolutely that there might be something more than meets the eye – more on this point later – but by the time he had finished the traditional arguments had been battered badly. No longer could one assume confidently that reason backs faith, justifying a belief not just in God but in the Christian God. Suddenly, intellectually, the Christian religion seemed as outmoded as it was socially irrelevant: harmful even.

Weathering the Storm

There were a number of reactions to this crisis. Some tried to carry on the old religion, in some way trying to stay with the old paradigm, as it were. Some tried to accept the critiques and to revise or recreate Christianity in a new or modern mode.

First, of the more conservative reactions, we have the rise of evangelical religion, marked in Britain in the eighteenth century by the coming of Methodism and in the nineteenth and twentieth centuries by numerous American sects (Marsden 1990, 1991). The main emphasis here has been on the claim that Jesus died to save all people from their sins (the belief known as Arminianism) rather than for just a select few (an extreme view probably accepted fully only by ultra-Calvinists). However, it is stressed that to be saved one must make a personal commitment to Christ. In other words, the emphasis is upon one's own free choice in making a commitment to Jesus, who will then shower his grace upon you. Anyone who has watched a televison programme featuring today's most famous evangelical preacher, Billy Graham, will know just how important is that personal commitment to Christ: the willingness to stand up and be counted. For the evangelical there is very much a personal relationship to God, and this is usually informed by His Word, that is, by the Bible.

Moral action is usually considered an important aspect of the evangelical's lifestyle. This is not performed in Pelagian fashion in order to buy one's way into the Kingdom of Heaven, but rather as a sign that one has received God's grace and wants to live in harmony with Him. Today, one often thinks of evangelicals as being socially conservative. In America particularly, with the inclination towards some fairly literal readings of the Bible, the Old Testament law rides high, with stern attitudes towards social justice, including such things as strong approval of capital punishment. Capitalism and anticommunism also are apparently the Lord's way. Historically, however, it is moral concern and action which are the evangelical's trademarks, and not necessarily the endorsement of one particular political philosophy. British evangelicals were leading forces in the fight against the worst excesses of industrialism and capitalism. Earlier, the evangelical circle of William Wilberforce (father of Samuel) had been notable in the fight against slavery.

An alternative approach, trying to ride out the criticisms of the last centuries, is one taken (at least until recently) by the Catholic Church and some more conservative branches of Protestantism. It is argued that the critics do not touch the essential tenets of Christian faith. As we shall see, notwithstanding the arguments of David Hume, the argument from design survived well past the eighteenth century, and there are many who would argue that it still has strength today. The same is true of such things as miracles. Here to a certain extent we find a defiance of the modern arguments, as the authority of the church is reasserted. The social structure of traditional Christianity is retained – note, for instance, the complete refusal of the Catholic Church to contemplate the ordination of women, a stand welcomed by many High Church Anglicans. The metaphysical structure is retained also. Catholics (unlike Protestants) still add to the list of saints, a process depending on the validation of miraculous signs as true marks of sainthood. Also, as with evangelicals, we find social and moral concerns, and these certainly reflect the theology of this particular cast on Christianity. The strong stand against abortion is a case in point.

Moving with the Wind

Other modern Christian approaches take the arguments of the Enlightenment very seriously indeed. Here it is argued that the truth must ride supreme and that one cannot let religion stand in its way. If science or philosophy or history tell us that religious claims may be challenged and can be shown wrong, then so be it. They must be revised or, if need be, rejected.

Most obviously, we have "liberalism." Crucial here was the thinking of the German theologian Friedrich Schleiermacher (1768–1834), who responded to Kantian criticisms of natural theology by stressing the significance of the subjective, of feeling, of "self-consciousness" for the religious person, and the way in which Christianity reduces to a sense of "absolute dependence" (Schleiermacher 1928). As liberalism develops, one finds that the claims of religion – for instance, about the sin of Adam or the resurrection of Jesus – are taken to be claims of the same epistemological standing as the claims of science or philosophy or history. If they do not stand the test against them, they must be rejected.

Given this stance, generally speaking what one finds from liberalism has been a steady retreat from just about all kinds of traditional theology. Little if anything emerges unscathed – whether it be the Old Testament, the New Testament, or beliefs added or embellished by Christians through the ages. Indeed, for many liberals what remains simply is the example of Jesus as a morally perfect man prepared to make the ultimate sacrifice as an example of unbounded love, and the moral code which is taken as embedded within the Christian message – a moral code which itself must adapt to our present-day understanding of such things as psychology and anthropology and biology. Lacking formal theology, liberal Christians turn for meaning to the ethical norms and demands of their faith. One finds them in the vanguard of many of the social movements of the past two centuries.

An alternative approach, one which accepts science as science and philosophy as philosophy, denies that this means one must gut Christianity of everything but a moral message. The influences here are epistemological, especially those of the existentialists like the nineteenth-century Danish philosopher Søren Kierkegaard, with his insistence that Christian belief requires a "leap of faith," going beyond the provable to the unknown: that in some sense, true religious belief demands this commitment. In the eyes of such believers, liberal (Protestant) Christianity seems unable to deal adequately with the problem of sin and of evil: a problem exacerbated and magnified in the last century in the most dreadful manner, even by (especially by) people who in many respects would seem the highest peak of humankind. The belief that religion reduces simply to a collection of ethical norms, without God's grace for sinners or the promise of salvation given freely by Him to us who are undeserving, seems a mockery in the modern age.

The most influential spokesman for this "neo-orthodoxy" was the Swiss theologian Karl Barth. A critic of natural theology – he found the God of the philosophers bore little relationship to the God of Christianity – Barth thought that ultimate reality is in some sense quite other and different from experience as we normally know it. It is not just strange but in some sense totally other. It is a question of different conceptual frameworks, in the strongest possible sense. We have our world: "[t]he world of human beings, and of time, and of things . . ." And then we have the unknown world: "the world of the Father, and of the Primal Creation, and of the Redemption." Different dimensions which are brought together for Barth the Christian through the person of Jesus Christ, our saviour: "[t]he point on the line of intersection at which the relation becomes observable and observed" (Barth 1933, 29). Moral behaviour is certainly not taken as unimportant, but – in a line going back through Luther to Augustine, and finally to Saint Paul – it is justification by grace which is the absolute centre of Christian belief.

Criterion of Comparison

The various positions just sketched are not the only options for a Christian today; but they are certainly some of the most important. And now that we have before us some of the major claims of Darwinism and some of the major claims of Christianity, we are ready to start a comparison. In the chapters that follow, I am not going to try for an exhaustive – and somewhat tedious – cross-check of every claim that every Darwinian has made against every claim that every Christian has ever made. I would not get past my first topic! I will not burden you at each point with my criterion of selection, but I should say that my aim has been to take on the more difficult and more fundamental issues. I am not a particularly heroic person, but an exercise like this would be nigh worthless if I always took the easiest option. We know before we start that the liberal Christian, as characterized above, is going to have little trouble with anything claimed in the name of science. So the views of people like this are going to get relatively little discussion in the pages that follow.

But equally, of course, if the discussion is to have real value, I must tackle problems involving beliefs which will be accepted or respected or acknowledged by today's Darwinians and today's Christians. So I will not be searching out esoteric beliefs held by some small band of believers, scientific or religious. My inclination therefore will be to compare a fairly strong form of Darwinism against fairly traditional forms of Christianity – Catholic and mainstream Protestant. As we will see, in my discussion of free will and original sin, if we can make some progress by comparing a natural-selection-incorporating scientific worldview with an Augustine-respecting religious worldview, then those in either camp who hold more moderate views (if we may so characterize them) can certainly build on the conclusions achieved. This would not necessarily have been the case had the starting points been the other ends of the scales. As is the nature of these things, I doubt that my approach will satisfy everyone. But you do know what it is and why I have taken it.

CHAPTER THREE

Origins

"Revealed religion" is the belief that comes through revelation or faith, directly or through the authority of others, and it is at this end of the scale that I shall start my examination of Darwinism versus Christianity. Obviously not every part of Christian belief, however central within the system, is of equal significance and importance to our inquiry. I doubt that evolutionism has much to say about the Trinity, for example. The question of origins, however, is crucial to Christianity, and this is (virtually by definition) a matter on which evolution has much to say. So let us begin there, specifically with the story of origins from the early chapters of the Old Testament.

Sacred Scripture

The Bible holds a special place in the faith of all Christians. There is a historical reason for this, going back to the Babylonian exile (587–538 B.C.) of the ancient Jews. Dispossessed of their lands and exiled far from home, they turned to their ancient traditions and writings as a means by which they could maintain their spiritual and national identity. The Torah, the law especially as given in the first five books (the Pentateuch) of the Bible, thus became a history and a prescription, as well as a promise of God's love and future concern: a salvation history (Farley and Hodgson 1994, 64–5). As such, these writings had a special status: given and inspired by God, they were more than merely contingent truths. They had to be true in every possible sense. And it was this attitude that was inherited and extended by early Christianity, as it developed its own

49

set of sacred writings: a set which included the Jewish testament, since it anticipated the coming of Christ and put the necessity of this coming into context.

At once we have our first point of potential conflict between the Darwinian and the Christian. Genesis comes complete with its own creation story, about God's having made Heaven and Earth and the denizens thereof in a mere six days – the first humans, Adam and Eve, being the culmination of the creative process. No absolute dates are given, but if one is sufficiently enterprising and imaginative – qualities possessed by some Christians in abundance – one can work back from the time of Jesus, using the detailed genealogies given in the Old Testament, and find that the creative process occurred about six thousand years ago, four thousand years before Christ. (Actually, Genesis has two creation stories, differing in details. In the first, Adam and Eve come together; in the second, Eve is made from Adam.)

On top of all of this, there are stories about subsequent events, most strikingly the universal flood which destroyed all but a very few humans and animals. Obviously none of this is compatible with evolutionism, which finds the origin of organisms in a developmental mechanism, with a 4.5-billion-year time span in which to do its work. And although evolutionism concedes that humans are a fairly recent arrival, it denies that they are necessarily the final creation – or indeed that creation through law is now finished. Notwithstanding the fact that the Flood belongs more to geology than to biology, there is still no place for this particular bottleneck through which all surviving organisms had to pass.

Given that the Bible has to be taken as the infallible word of God, how committed is the Christian to reading Genesis in such an absolutely literal sense? Must it be in such a sense as to bring Darwinian evolutionary theory into unresolvable conflict with the Bible? Let me start historically and then move to analytic discussion.

Catholicism

I begin with the Catholic faith, and by now you will not be surprised to find that my first point of reference is Saint Augustine of Hippo, back in the fourth and fifth centuries of the Christian era. Raised a Christian, Augustine later joined the Manichees. This sect – believing that matter is inherently bad – thought that the earth is a battleground between a force for good and a force for evil, that Jesus, although a prophet was not divine, and that the Old Testament is pernicious nonsense. When he converted back to the true faith, Augustine realized that a major task must be the fighting of the Manichean heresy. Were one to insist on an absolutely literal reading of Genesis, one would only be giving comfort to the enemy, who would seize on the inconsistencies both within the Bible and between the Bible and commonly accepted scientific beliefs. "It is a disgraceful and dangerous thing for an infidel to hear a Christian, presumably giving the meaning of Holy Scripture, talking nonsense on these topics; and we should take all means to prevent such an embarrassing situation, in which people show up vast ignorance in a Christian and laugh it to scorn" (Augustine 1982, 42–43, I, 19, 39).

Building on the thinking of others (especially Origen) who had realized that there are dangers lurking here, Augustine developed the influential thesis that Moses (the supposed author of Genesis) had to write in metaphorical or allegorical form, because the ancient Jews were untutored in science. What Moses said was not false, but not necessarily the literal truth. As a matter of practice, we today should start by accepting the Bible at literal face value. But if the Bible taken literally is clearly false, then interpretation is permissible. Although Augustine was certainly no evolutionist in the sense that we understand the term, as it happens he himself seems to have favoured at least a developmental view of organic creation. In his theology, it was crucial that God should have created everything in one move directly out of nothing: the conception and wish and the creation were the same. Augustine believed therefore that organic forms were created potentially in a kind of seed form, and realized actually when the conditions were right – when the seas appeared, for instance.

One difficulty on the Augustinian system is the question of just how firmly demonstrated a science must be before it overrules literal scripture. If one thinks of the Augustinian position as demanding (in the language of the courts) that scientific claims conflicting with scripture be proved correct beyond a reasonable doubt before they are accepted, then one might yearn for a weaker criterion. Perhaps science should be accepted if it is more probable than not. There are hints of this even in Augustine, but one has to wait until the scientific revolution before the position is articulated fully, notably by Galileo (McMullin 1981). His claim was that truth cannot be opposed to truth, and that since the truths of science as well as the truths of revelation both come from God, there cannot be genuine conflict. Hence, since in essence religion and science are evenly balanced, there should be no a priori presumption in favour of religion, and science should be allowed to get on with its job.

Nothing physical that sense experience sets before our eyes, or that necessary demonstrations prove to us, should be called in question, not to say condemned, because of biblical passages that have an apparently different meaning. Scriptural statements are not bound by rules as strict as natural events, and God is not less excellently revealed in these events than in the sacred propositions of the Bible. ("Letter to the Grand Duchess Christina," in Galileo 1957, 182–3)

Notoriously, Galileo failed to carry the authorities along with his way of thinking. But his was the time of the Counter-Reformation, noted already as the time when the church was in a desperate battle with the Protestants and could ill afford any slackening of its position. With the successes of science in the past four centuries, it has become apparent that the more moderate approach to the science/Bible tension is demanded, and indeed since the end of the nineteenth century this Galilean moderate approach seems to have been the church's position. (The pertinent authority is the encyclical *Providentissimus Deus* of Pope Leo XIII, 1893.) Today, the highest authorities now accept that evolution – even Darwinian evolution in some respects – is sufficiently well established that one might licitly hold to it. "The theory of evolution is more than a hypothesis. It is indeed remarkable that this theory has been progressively accepted by researchers, following a series of discoveries in various fields of knowledge. The convergence, neither sought nor fabricated, of the results of work that was conducted independently is in itself a significant argument in favour of this theory" (John Paul II 1997, 382, translation corrected).

Protestantism

What about the Protestant position on the creation story of Genesis? Here one might expect more tension, especially from the mainstream branches. The mark of the great reformers was, above all, a renewed emphasis on the Bible, as one searches for understanding of God's grace working in this imperfect world of ours. It was no mere chance that it was Luther who was the greatest of all of the translators of the Bible into the vernacular. Yet, historically and conceptually, one must take care. As the

Reformation hardened into place around 1600, one could find many who were taking a strict line on the literal truth of all of the Bible. "No error, even in unimportant matters, no defect of memory, not to say untruth, can have any place in all the Sacred Scriptures" (Calovius, quoted in Dillenberger 1960, 97). But Luther and Calvin themselves took a more tolerant line on interpretation. For them, what counted was not a lifeless literalness; rather, it was a faith based on Jesus Christ. All else was subsidiary, to such an extent that we find Luther positively contemptuous of those parts of the New Testament that he did not consider sufficiently in line with his own beliefs. Notoriously, he referred to the Epistle of James as "right strawy stuff." Calvin was more concerned to stay with a literal reading of the Bible, but he too realized that some interpretative work was needed. To this end, he introduced his famous doctrine of "accommodation," which recognized that the Bible is sometimes written in such a form as to make itself intelligible to scientifically untutored folk who would not have followed sophisticated discourse.

Moses wrote in a popular style things which, without instruction, all ordinary persons endued with common sense, are able to understand; but astronomers investigate with great labour whatever the sagacity of the human mind can comprehend. Nevertheless, this study is not to be reprobated, nor this science to be condemned, because some frantic persons are wont boldly to reject whatever is unknown to them. For astronomy is not only pleasant, but also very useful to be known: it cannot be denied that this art unfolds the admirable wisdom of God. . . . Nor did Moses truly wish to withdraw us from this pursuit in omitting such things as are peculiar to the art; but because he was ordained a teacher as well of the unlearned and rude as of the learned, he could not otherwise fulfil his office than by descending to this grosser method of instruction. . . . Moses, therefore, rather adapts his discourse to common usage. (Calvin 1847–50, 1, 86–7)

Luther and Calvin both accepted Ptolemaic astronomy, more because it was the sensible position of their day than because either was desperately concerned to defend every last literal word of Scripture. In fact, in the time of the sixteenth century after Copernicus (1543–1600), there were only ten full-blooded Copernicans: seven were Protestant and three were Catholic (Westman 1986). Among the Protestants were Osiander, who wrote the introduction to the *Revolutionibus* (admittedly with an instrumentalist interpretation, denying that one must take heliocentrism literally), and Georg Joachim Rhetticus, a man who was in the circle of Luther's fellow reformer Melanchthon and who was allowed by Copernicus to publish a preliminary version of the heliocentric theory (*Narratio Prima,* 1540).

As we get into the seventeenth century and the rise of modern science, accommodation is in full swing. The exact connection between Protestantism – Calvinism in particular – and science is much debated, but it was clear that no literalistic reading of Genesis was going to stand in the way of empirical discovery. "Truth cannot be opposed to truth." What then of the question of evolution and Genesis? By the time that we reach the nineteenth century, scientists have realized that the Earth must be far older than a mere few thousand years. The industrial revolution forced this fact upon people. The need to dig for coal and for minerals, the need to drive channels and tunnels through solid rock to build canals, the search for building materials and more, made geology a vital and necessary science – and with its development came the realization that only by postulating long periods of time could one hope to bring any ordered understanding of the Earth's past.

Admittedly, even into the nineteenth century respectable scientists went in for a certain amount of "Biblical geology," trying to find confirmation of specific scriptural claims in the rocks (Millhauser 1954). The Reverend William Buckland, professor of geology at the University of Oxford, published (in 1823) a work supposedly giving evidence of Noah's flood, but this kind of work was really out of fashion (Figure 7). The competing schools of the 1830s, the uniformitarian geology of Charles Lyell and the catastrophism of most others, including the ordained professorial scientists at Oxford and Cambridge, both denied Genesis taken literally. The former supposed massive amounts of time and the slow-but-certain action of forces like those around us today (wind, rain, frost, and so forth). The latter, derived from the geological speculations of the great French biologist Georges Cuvier (1813), supposed that the earth is cooling from a molten state and that every now and then a huge upheaval destroys what had existed before, leaving the slate clean for new creations. It is true that the catastrophists could be found worrying whether the six days of creation referred to six long periods of time, or if there was a long unmentioned period of time separating various of the (twenty-four-hour) days, but this was religion fitting itself around science. When Philip Gosse in *Omphalos* (1857) suggested that perhaps the fossils were

7. One of the illustrations Buckland gave (in his *Reliquiae Diluvianae*) purporting to prove the existence of a universal flash flood. Skeleton G is of a rhinoceros, washed down into the cave and enclosed in diluvial rubble (E). That there were bones to reconstruct only one rhinoceros, Buckland claimed, proved the extreme rapidity of the flood. (Buckland, W. *Reliquiae Diluvianae*, 1823.)

artifacts put in place by God to test our faith, he was ignored as outside the bounds of science: and of true religion, for that matter.

Thus, by the time Darwin arrived with the *Origin of Species,* Protestants had already dealt with the Genesis problem (Ruse 1975b, 1979a). Those early books were simply not to be taken as literal descriptions of scientific reality. There was still much tension about what we have seen called higher criticism, where biblical scholars treated the texts as one would secular documents, giving naturalistic accounts of the Genesis stories; but the intellectual world was well on the path to modern-day thinking, where the first creation story of Genesis (dated from 600 B.C. and the Babylonian exile) is seen as an affirmation of God's power and caring for a people dispossessed, and the second creation story of Genesis (dated from 1000 B.C. and the time of King David) is seen as an articulation of the place and role of a king (symbolized by Adam) with respect to his people (symbolized by the rest of creation) (Bergant and Stuhlmueller 1985).

After the *Origin*

Why then was there so much religious controversy when the *Origin* was published? There are two answers. The first is that there was not so much religious controversy! I noted in the Prologue that the most famous clash – that between Wilberforce and Huxley – is mainly a later invention of the triumphant evolutionists, who wanted to show that the dragons they had slain were larger than they truly were in real life. The reality is that, even though people pulled back from selection, evolution as fact (and as something demanding a theory) was accepted very rapidly. Listen to a future archbishop of Canterbury: "Even those who contend for the literal interpretation of this part of the Bible will generally admit, that the purpose of the revelation is not to teach Science at all. It is to teach great spiritual and moral lessons, and it takes the facts of nature as they appear to ordinary people" (Temple 1884,180–1).

The second point to be made about the post-*Origin* period is that the bulk of such controversy as there undoubtedly was did not centre on literalistic readings of Genesis (Moore 1979). There were other factors – factors which we shall consider later – like the problem of design and the place of immortal souls, which loomed much larger. Sophisticated churchmen, then and now, just did not think the creation stories of Genesis or related claims, like that about the Flood, to be significant factors in considering the truth of evolution, or of any theory of causes.

All of this is to ignore the current controversy about "creation science." If everything was as warm and friendly as I am suggesting, why then did I have to fly south to Arkansas in 1981? Why is it that there are still today articulate evangelical Christians who believe that one must revert to a strict biblically based account of origins: six days of creation, humans last, universal flood, and much else, even including for many a very short history for the Earth? For technical legal reasons – fear of falling afoul of the U.S. Constitution's separation of church and state – these people usually do not flaunt the biblical underpinning of their thinking, but it is there nevertheless. For such believers, Darwinism and Christianity part ways right at the beginning. There is simply no way that a Darwinian can be a Christian.

The people to whom I now refer are not traditional Christians: at least, they do not stand in the Christian tradition – Catholic or Protestant – of reconciling science and religion. But why do they exist at all? There are a number of reasons, both at the individual level and at the broader social level (Marsden 1980; Numbers 1992; Conkin 1998). Everything starts in the nineteenth century, in America particularly, with traditional Protestantism taking on a much more fervent evangelical cast, with personal attestations of faith and claims that sincerity rather than intellect and training are the true marks of the believer. One saw then the rise of transformed versions of older sects and of new forms of belief and sects – near-cults in many cases. One mark of this individual based religiosity was a turn to the Bible and a renewed interest in interpretation, often of those passages which seem to tell of the future: naturally of much personal interest to those conquering a new land.

Revelation, with its forecast of Armageddon, came in for much sympathetic scrutiny, and many believers found the tales of Genesis – creation and then a flood – direct mirror confirmation of what was to happen in the future. Alpha and Omega, as it were. One such group of "dispensationalists" (believers in eras culminating in catastrophes) were the followers of Mary Ellen White, the Seventh Day Adventists. At the turn of the century, one enthusiast, a Canadian travelling salesman named George McCready Price, devised a whole system of geology which supposedly gives complete scientific backing to Genesis taken literally. Normally, such ideas – Price had no scientific training whatsoever – would have received short shrift, even from the most enthusiastic believers; but as we come into this century the extreme evangelical movement was gaining strength rather than fading. Particularly in the South, the loss of the Civil War and the subsequent social upheavals, the coming of industrialism, the massive immigrations from Europe of Catholics and then Jews, the liberal turn of many of the older, more conventional churches, the successful fight against perceived evils like alcohol, and much more gave impetus and strength to those who wanted to interpret the Bible literally. And there were many who found views like those of Price congenial.

Thus we get the rise of that particular version of Biblical literalism known as "fundamentalism," culminating in the 1920s in the notorious

Scopes trial. It is also no real surprise that fundamentalism fell back in retreat when America laughed at its pretensions and naivete. Fell back, but was not vanquished. I have told of how the Russian success with sputnik in 1957 set in motion a chain of events – especially through the production of new science textbooks – which led to the renewal of the creationist movement. The hour of need found the men of resource. Needing to refurbish the Flood geology of McCready Price, two men – John Whitcomb, a biblical scholar, and Henry Morris, a hydraulic engineer – stepped forward in 1961 with their jointly authored *Genesis Flood*. This set the scene leading to Arkansas and beyond.

Unacceptable Science

So much for the history. We can now see that traditional Christianity does not demand a literalistic reading of Genesis, and yet we now know why it is that a movement sprung up insisting on a literal, anti-Darwinian reading of the sacred text. At one level, given the scope of the discussion, our work is finished. We are not asking the question, Is Darwinism true? Rather, having assumed the truth of (some version of) Darwinism, we are asking, Can a Darwinian be a Christian? The answer is very obviously (with respect to Genesis) that one can be, whether one be a Catholic or a Protestant. In fact, most Darwinians – and here I speak of all shades, from ultras like Dawkins through qualifiers like Gould – would argue that the evidence for evolution and for some significant role for selection is sufficiently strong that Christians ought to be Darwinians. Our powers of sense and of reason are given to us by God – they are crucially involved in what it means to say that humans are made in God's image – and to turn our back on such firmly established science is theologically unacceptable.

Not just theologically unacceptable, of course, but also threatening to science, in the sense that if one's Christianity is such that one can disregard such strong science, then one's religion allows a luxury of interpretation and denial that no scientist (including the Darwinian) could permit without compromising important standards. Indeed, it is worrisome to think that – because of a literal reading of the Bible – we could have the live option of rejecting such established science as Darwinism. Consider the position of Alvin Plantinga, a Calvinist and America's most distinguished living philosopher of religion, who writes even of the hypothesis that the Earth is very old (something he accepts) let alone of the hypothesis of evolution (something he does not accept), that "[a] sensible person might be convinced, after careful and prayerful study of the Scriptures, that what the Lord teaches there implies that this evidence is misleading and that as a matter of fact the Earth really *is* very young. So far as I can see, there is nothing to rule this out as automatically pathological, or irrational or irresponsible or stupid" (Plantinga 1991b, 15).

At the risk of sounding intolerant, any scientist – including any Darwinian – has to insist that there comes a point at which discussion is closed. No sensible person could or should possibly be convinced, after no matter how much careful and prayerful study of the Scriptures, that the Earth is the centre of the universe with the sun going around it in a circle. Such a belief is irresponsible and stupid, and if one's religion allows this, then the scientist (including the Darwinian) has to reject the religion. It is on a par with native beliefs that the Great Spirit will protect them from the white man's bullets. And here is the crunch. Darwinians today think their theory sufficiently well established that, if Christianity is a religion which would even allow the reasonable possibility of Darwinism's rejection on grounds of conflict with literal readings of Scripture (not that one would necessarily oneself reject it, but that one would think it reasonable for someone to reject it), then Christianity itself ought to be rejected.

Note that this opening of such a possibility is not sanctioned or approved or endorsed by traditional Catholicism or Protestantism. It is based on an idiosyncratic, twentieth-century, American reading of the science/religion relationship. Of course, Plantinga might reply that Darwinism does not fall into the same category of well-established scientific theory as Copernicanism. But that is another point, one not at issue here. The assumption we made at the beginning of our discussion is that Darwinism in some form is well taken. Our exercise is the exploring of the implications of that assumption for Christian faith. Although, particularly given that it is Plantinga against whom I am now arguing, I must stress again that disagreements between evolutionists, even between Dawkins and Gould – the two loudest and most visible disputants about the evolutionary process – pale beside their agreements and their shared appreciation of the extent to which today we build on Darwin, while at the same time knowing so much more than Darwin did.

Let me emphasize this point. Creationists often like to trade on an ambiguity between evolution as fact and evolution as path and cause, suggesting that because there is debate about the latter there is doubt about the former. Plantinga is a past master at this gambit. But it is an ambiguity, and it is being used falsely. (See Plantinga 1991b; for more details on this confusion, see also Scott 1996, 1997.) No one today thinks that evolution as fact is any less secure than it was for Darwin. For life scientists, that really is a given, like the movement of the Earth. So if one's Christianity – which is apparently true of Plantinga's Christianity – is such that it is reasonable to allow someone to deny Earth's long evolutionary history, then Darwinism and Christianity are incompatible. And this is before we get to causes, which will only strengthen the division. Today, no one denies our debt to Darwin or that our shared knowledge of path and cause is much greater than it was for Darwin. If nothing else, genetics – first Mendelian and then molecular – has transformed our understanding, and given a whole new appreciation of selection. I am not arguing this point or the underlying strengths of the beliefs held by today's evolutionists: I am stating categorically that this is where people across the spectrum do stand today. For all the doubts and arguments and claims and counterclaims, with very few exceptions, natural selection is seen as a significant cause of evolutionary change.

But I stress again that Plantinga's Christianity is not traditional Christianity. With respect to the issues we are considering here, Darwinism can be accepted by the Christian: allegorical readings of Genesis are permitted and in such a case as this perhaps even mandatory. Which conclusion now raises a different worry. Have we now the thin end of a wedge which ends with the general claim that Christianity must give way before every putative scientific move? Must Christianity today defer before every speculation or hypothesis which is made in the name of science? What about the very origin of life itself? Surely here we have an issue on which Darwinism and Christianity part company. Let us see if this is indeed so.

Origin of Life

"Is there any point to which you would wish to draw my attention?"
"To the curious incident of the dog in the night-time."
"The dog did nothing in the night-time."
"That was the curious incident," remarked Sherlock Holmes.

In Conan Doyle's brilliant short story "Silver Blaze," a nonevent is the crucial clue to the discovery of the fate of the stolen racehorse and the murder of its groom. The dog did nothing because the intruder was no stranger, but rather the familiar groom himself, who was justly killed by the racehorse as he attempted to nobble it before a big race. In Charles Darwin's brilliant long story, *On the Origin of Species,* a nonevent is the crucial clue to the discovery of Darwin's thinking on the origin of life itself. Darwin simply does not mention the topic at all, remarking only that life has evolved from one or a few forms.

The omission was quite deliberate. The *Origin* was published just at the time when the French scientist Louis Pasteur was driving the final nail into the coffin of spontaneous generation, the belief that life comes naturally and in one leap from nonlife: worms out of mud, and that sort of thing (Farley 1977). Darwin knew full well that evolutionary speculations lead to thoughts of ultimate origins – Lamarck had all sorts of ideas about how heat and electricity produce lower forms from slime and dirt – and Darwin (a very knowledgeable and skilled scientist) knew full well that such speculations would bring down scorn and disfavour on his general theorizing about evolution. On the sensible policy that sometimes it is best to say nothing, nothing is precisely what he did say. And very successful he was in his strategy, for the origin of life question was not a major factor in the critical discussion of the *Origin.*

But it was an issue. Despite his silence, Darwin knew that the original appearance of life is something which presses upon the evolutionist. Indeed, he himself did later speculate (privately in a letter) about how proteins might have been formed from the action of lightning and so forth on "some warm little pond, with all sorts of ammonia and phosphoric salts" (Darwin 1887, 3, 18). Yet, although Darwin's supporters and followers, notably Huxley, were far less reticent on the subject, there was no real progress until the 1920s and 1930s, when renewed interest in the subject started to produce new ideas and discoveries. Today, no textbook or general discussion on evolution would be complete without a chapter on the subject. You may not find a discussion of ultimate origins in the *Origin of Species;* you will find a discussion in Richard Dawkins's *The Blind Watchmaker.*

But is there not here a point of particular conflict with the Christian? Is not the case for a natural origin of life so ludicrously thin that one can hold it only through some sort of blind faith in science; and if one were to take seriously the demands of religion, would not one have to reject it and opt rather for a Genesis-inspired divine origin of life? Take Augustine's criteria for resolving science/religion conflicts. Surely, no one could say that scientific origin studies today are so secure that one can legitimately hold to them against Genesis. Plantinga speaks of hypotheses about the origin of life as "for the most part mere arrogant bluster," adding that "given our present state of knowledge, I believe it is vastly less probable, on our present evidence, than is its denial" (Plantinga 1991b, 20). Hence, if as a Darwinian one feels obliged to hold to the legitimacy and validity of such studies – and contemporary Darwinism does seem to demand this – then one is going to fly in the face of what one ought to do as a good Christian. In other words, in this respect, a Darwinian cannot be a genuine Christian.

Are things really as bad as this? We must look at the state of the science.

The Science of Origins

Today's starting point is the thinking of the Englishman J. B. S. Haldane (1929) and the Russian A. I. Oparin (1928, 1957). Marxists in later life but not when they first had these ideas, they independently proposed a similar, tentative, sequential route that originating life might have taken. First, organic molecules are made from inorganic molecules, by natural processes like lightning. This is not the old-fashioned spontaneous generation, because we are talking now only of the building blocks of life and not of life itself. In the second stage, such building blocks are joined together into the macromolecular chains (like proteins and amino acids) that make up living organisms here on earth. Finally, third, somehow these macromolecular chains start to replicate and feed off the "prebiotic soup," which is the state of ponds and so forth as the result of the first stage of evolution. After this, life is on its way, although clearly there are going to be more stages, such as those responsible for the formation of cells, the self-contained internally functioning components of living beings (Freeman and Herron 1998).

The possibility of the first stage of the Oparin-Haldane hypothesis is now strongly confirmed. Well known are the experiments starting with those of Stanley Miller and Harold Urey in Chicago in the 1950s, which set up apparatuses holding inorganic chemicals, subjected them to heat and electric shock – simulating what they thought would be the state of a young earth – and were able to obtain organic molecules (amino acids, the components of proteins) quite naturally and very rapidly (Miller 1953, 1992). Since then, many other organic molecules have been obtained, including the components of nucleic acids (which carry the genetic information needed by organisms). The second stage of the hypothesis has also started to yield to experimental (and theoretical) attack. In particular, it is thought that naturally occurring clays, compounds including aluminum and silicate, might be crucial catalysts (Ferris et al. 1996). Organic molecules stick to the clays and can assemble spontaneously into chains. After a while, these macromolecules become sufficiently stable in themselves that the clays are no longer necessary; these latter can get washed or leached away, leaving the biological molecules themselves intact – able, indeed, to add more units to their ends and continue growing in length.

The third stage of the Oparin-Haldane hypothesis – when the molecules become self-replicating – is still more tentative. Some students of the subject think that the clays continue to play a major and significant role. Clay components can crystallize, and such crystals can serve as the seeds for further crystals. Most importantly, sometimes crystals contain errors or imperfections, and these can be transmitted to offspring crystals. Perhaps, therefore, what happened was that the biological molecules piggybacked on the clay crystals, themselves also replicating and picking up errors – kinds of proto-mutations. Eventually, the macromolecules on their own developed the ability to reproduce, and the clays were dropped as redundant (Cairns-Smith 1982, 1986).

Other students favour hypotheses according to which the macromolecules themselves unaided start replicating. Ribonucleic acid (RNA) – the molecule which in most organisms reads the genetic information from the carrier, deoxyribonucleic acid (DNA), and then serves as the template for making proteins – is a favoured candidate for the key molecule in this process. This is both because independent studies on the classification (taxonomy) of living organisms point back to the key shared role of this substance (RNA), and because in some organisms today RNA alone carries the genetic information and at the same time acts as the catalyst to produce the protein components of the cell (James and Ellington 1995).

At this point, no one has actually confirmed that RNA can start self-replicating, nor has such an event been reproduced experimentally. But already major strides are being made in this direction. Biochemists can get RNA molecules to copy themselves in the right (admittedly artificial) media, can get the molecules to add to their lengths, can show how there will be variation in these additions, and can show how this can be crucial in such biological functions as catalysing other molecules. We have a kind of proto-mutation and a kind of proto–natural selection. At the moment, the hand of human design and intention hangs heavily over everything, but work is going forward rapidly to create conditions in which molecules can make the right and needed steps without constant outside help. When that happens, as one researcher has put it, "the dreaming stops and the fun begins," as one looks to see if such a replicating RNA molecule could make a DNA molecule, and how it might function both with respect to producing new cell components and with respect to transmitting information from one generation to the next (Fox 1988; Bartel and Szostak 1993; Ekland et al. 1995; Ekland and Bartel 1996).

Christian Responses

You do not need to accept all of the interpretations that the scientists put on this work. Indeed, the scientists themselves acknowledge that they have not yet arrived at a solution to life's origin. One prominent researcher, using the metaphor of a detective story, writes: "We must conclude that we have identified some important suspects and, in each case, we have some ideas about the method they might have used. However, we are very far from knowing whodunit" (Orgel 1998, 495). But if anything the problem presents an embarrassment of riches – we have identified many different possible solutions to key steps – rather than a paucity of ideas on the subject. There is, for instance, some considerable controversy over the actual conditions which really existed when the first significant organic molecules were formed. It may be that the required hydrogen was not freely available (what scientists call a "reducing" atmosphere). Perhaps, therefore, the scene of the crime was not so much Darwin's warm little pond exposed to the air (as it then was) but deep sea vents throwing up superheated water, rich in minerals, reacting with the cold sea water. Time will tell which of these and similar various options will prove best supported. The point is that disagreement among the scientists does not preclude your need to acknowledge the existence of all of this work, undoubtedly speculative and tentative though much of it remains. You must take it seriously. To say with Plantinga that all of this work is "for the most part mere arrogant bluster" is just plain silly. (Fry

1999 has a full discussion of both the pertinent science and the underlying philosophical presuppositions.)

What then of the Christian who does want to take science seriously, including this extension (as we might call it) of Darwinism (for note that the biochemists are now starting to think in terms of the ways in which selection might take over the process)? If one insisted on taking a rigid Augustinian line, then even now one might feel compelled to argue that the origin of life is a mystery and that as such, given sacred Scripture, one ought to prefer an answer in terms of a special intervention by God. I confess that I am of two minds about this move. My tolerant inclination is to let it be. I would not say – as I would of the person who denied Copernicanism, and as Darwinians would say of the person who denied the basic aspects of Darwinism – that one stood in danger of the unchristian act of denying one's God-given powers of sense and reason. One may disagree with a person so sceptical about origin-of-life studies, but one could retain respect for his or her Christian integrity. My underlying feeling, however, is that – even as a general point – given the massive successes of science in the past four centuries, a rigid Augustinian line is no longer appropriate. Augustine was not thinking within or against the history of a successful scientific culture. Even if it had still been a matter for debate at the time of Galileo, I rather doubt that, were he alive today, Augustine himself would insist on the preemptive truth of a literal reading of the Bible. If he felt the need of a policy to deal with the Manichees, how much more would he feel the need of a policy to deal with modern science?

Certainly, with respect to the problem of life's origins, even the neo-Augustinian ought to pause and wonder if so conservative a position (as a categorical appeal to miracle for explanation) is really merited on Christian grounds. And the same holds very much more for the Catholic who takes Galileo's line, as well as the Protestant who takes seriously some version of Calvin's accommodation. Consider matters both ways. From the perspective of science, this seems about the silliest of all possible times to give up on the origin-of-life problem. One has had half a century of molecular biology which has taught us an incredible amount about the cell and its constituents. On top of this, one has had massive advances in related pertinent sciences: the Big Bang theory in astronomy and plate tectonics in geology, for instance. And still the discoveries pour forth. Now, when the tools are really starting to lie at hand, is not the moment to declare defeat. It is true that the actual puzzle of life's origin is very tough, but already one has a puzzle where a great many links have been completed or suggested, and where there is fast movement forward on others. This is not an area which has stood still for many years, without hope or prospect of advance – an area where one might start to think that science has come to a dead end. It is rather a good place for the scientist to work: the sort of place where talk of Nobel Prizes is not inappropriate. It may not pan out quite as people think; exciting science rarely does. But the odds are that something will pan out in some way before too long. "The only certainty is that there will be a rational solution" (Orgel 1998, 495).

This being so, it is not unchristian to accept this, expecting that scientists will make the key discoveries and breakthroughs: to declare, indeed, that one now thinks that life came naturally from nonlife. Even if one cannot go this far, the Christian should be counselled to cover the bets. It would be very foolish to declare too definitively on the impossibility that life will, at some point, be shown to be the natural outcome of processes starting with the inorganic. At the very least, prudence would dictate caution. It would be best to declare oneself an agnostic on the question of the origin of life. Perhaps science will explain all, perhaps not. Which position, I take it, is roughly that of the popular writer on science/religion issues Paul Davies (1999), although one should caution that he is not – as we have assumed as a basis for discussion – a fully committed Darwinian. He denies that life's beginning was supernatural, but he does seem to want to leave open the possibility that there was a little bit of something more. "I do believe that we live in a bio-friendly universe of a stunningly ingenious character" (20). He is looking for "new philosophical principles" which will have "immense philosophical ramifications," and seems to think that information will be important.

But, however one locates oneself on this issue, even the doubter would be advised to agree that – considering matters now from the perspective of the Christian – nothing terribly important rests on this scientific matter, either way. This point should be stressed. Prudence aside, from the Christian viewpoint a certain theological openness to the success of science in this domain is warranted. *Ex hypothesi,* we have taken it that Darwinism in the general sense is sufficiently well taken that one should, qua Christian, accept it. Hence, one is already committed to a significant allegorical reading of the early chapters of Genesis. One agrees that life came developmentally from primitive forms to the marvellous array of complexity that surrounds us today, and that such evolution is likewise revealed in the fossil record. One agrees that God could and did do all of this, just as one agrees that God could and does guide the growth of every individual from the fertilized cell to the fully grown adult form. This being so, why then cavil at the original first step?

Apart from anything else, in the light of modern science, theologically does anything so very significant rest on this, any more? One is addressing the question of life itself, rather than the far more significant theological issue of the distinctive status of humans. Today no serious biologist thinks, however life began, that there are special life forces – what the French philosopher Henri Bergson (1907) called *élans vitaux* – which animate organisms, making them alive. There is no physical evidence for the existence of such forces. They are not akin to unseen theoretical entities like electrons because, even if one supposes that they do exist, they do not seem to do or explain anything very much. Rather, life is all a question of organization and getting on with the job. I shall say more about these and related matters in the next chapter. The point is that accepting a natural origin of life is not a case of giving up because one cannot hold the hordes back. Rather, one is realizing how far one must interpret the writings of the ancient Jews in the light of modern science. One now sees the extent to which such an interpretation must extend. There is nothing wrong or even very surprising about this. The fact that we may not much care for the full implications is beside the point. We do not always care to put aside childish things. As the apostle Paul warned us, such is the nature of growth.

I conclude, both from the viewpoint of science and from the viewpoint of religion, that if one's understanding of Darwinism does include a natural evolution of life from nonlife, there is no reason to think that this now makes Christian belief impossible.

CHAPTER FOUR

Humans

It was Thomas Henry Huxley, the master of the public forum, who set the agenda on the "monkey question." In his little book *Man's Place in Nature* (1863), he argued that the anatomical similarities between humans and apes provide overwhelming evidence of our primate evolutionary origins. It is true that there are major differences between humans and the closest apes – the gorillas, the chimpanzees, and the orangutans – but the differences between us and them are less significant than are the differences between them and the lowest and most primitive monkeys.

There was some fossil evidence also. Thanks to the patient efforts of the Frenchman Boucher de Perthes, human origins were being pushed back in time, because our remains were being uncovered along with specimens of now-extinct organisms (Oakley 1964; Johanson and Edey 1981). Then, bones of an apparently primitive form of human were unearthed from the deposits in the Neanderthal region in Germany. Not that these were necessarily the much-desired and sought-for "missing link": Huxley himself, for instance, decided that Neanderthal man was but a subspecies of *Homo sapiens,* rather than a new species in its own right. But the fossil discoveries did point in the right direction: a direction that was taken by the Dutch doctor Eugene Dubois at the end of the century when, in the East Indies, he discovered remains of the first unambiguously primitive humanlike (hominid) specimen (formerly "Java man," *Pithecanthropus,* now assigned to *Homo erectus*).

Human Evolution

The Descent of Man (1871), the work that Darwin did eventually write on humankind, is (to be candid) somewhat derivative. As always less interested in questions of phylogeny than in those of cause, he did make it clear that his favoured birthplace for humankind was Africa rather than the East. In this century, there have been massive finds from that then-mysterious continent, confirming Darwin's beliefs. Jumping now to the present day, human origins provide one of the best of all worked-out lineages. We start back about four million years ago, with small creatures about half our height: *Australopithecus afarensis.* Then we go through later specimens of the same genus, *Australopithecus,* moving next to the genus *Homo:* first *Homo erectus,* through *Homo habilis,* and then (depending on how you date the change) between 500,000 and a million years ago we get the appearance of *Homo sapiens* (Pilbeam 1984; Wolpoff and Caspari 1997) (Figure 8).

Thanks to this detailed path, we can now answer definitively a question which puzzled Darwin and his contemporaries: Which came first, walking or the big brain? *Australopithecus afarensis* had a small brain about the size of an ape's, 500 cc; and then through the successive species up to *Homo sapiens* we get the evolution of increased size to the present 1200 cc. *A. afarensis,* however, had already come down from the trees and had evolved an upright stance, being a fully functional walker – although significantly and interestingly, not quite as good as we humans. Our ancestors had stood themselves on two feet rather than four before they started to upgrade their brain power.

The fossil record for the years before *A. afarensis* is not good, and one turns here to molecular traces for help. Although at first it shocked people considerably, it seems now to be firmly established that it was only five or six million years ago that we split off from the (ancestral) lines leading to the gorillas and chimpanzees, our closest relatives; and indeed, we humans may well be more closely related to chimpanzees, than chimpanzees are to gorillas. Precisely why we humans evolved as we did – after we came down from the trees and started walking upright – is still a matter of much debate and dissension, although no one doubts that selection played a major role. The enthusiasts for punctuated equilibria suggest that the fossil record fits the jerky pattern one would expect if their theory were true, but ultra-Darwinians argue that it better fits a

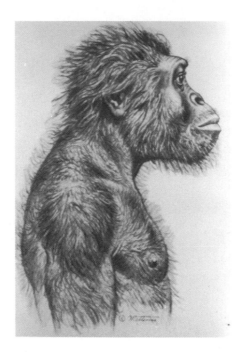

8. *Australopithecus afarensis.* (Johanson, D., and M. Edey. *Lucy,* 1980.)

path marked by gradualism. To a certain extent, it is a matter of what you mean by "jerky" and what you mean by "gradual" (Eldredge and Tattersall 1982; Isaac 1983).

The Harvard biologist and keen Darwinian Edward O. Wilson (1975) proposes what he calls a theory of "autocatalytic" evolution, where at points or thresholds one gets a positive feedback and evolution goes very rapidly. He suggests that in human evolution this could have happened twice. First, when we moved upright: either be on all fours or on two legs, but not in between, so there is massive selection pressure not to delay the evolution from one to the other. Second, when we developed brains: they are so expensive to produce that one needs really big ones or their benefits do not outweigh their costs. Here we would have irregular rates of evolution for purely Darwinian reasons: perhaps a selection-fuelled, punctuated equilibria pattern.

Although the speculation level has always been high when it comes to the causal factors behind human evolution, there are increasing numbers of findings and techniques which are lifting speculation from a metaphysical to a genuinely scientific level. For instance, studies on teeth can be very revealing about diet, and combined with fossil discoveries of animal remains along with human remains they can give major insights into our ancestors' feeding practices and their social habits. It does seem clear that much of human evolution was marked by scavenging behaviour – humans were the jackals of the primate world – combined with major moves towards sociality. It is very obvious that humans do not generally have the adaptations that predators like lions and tigers – and even chimpanzees – have as a matter of course. We are not that strong or fast. We succeed because we can think and work together, after a fashion at least. Scavenging may not be very dignified, but it is a good feeding strategy for such animals as we became.

Clearly, a major factor in sophisticated sociality is the development and use of language. Animals can make sounds and communicate: as we now know, sometimes in remarkably intricate and clever ways, but nothing to compare to humans. Darwinians today think that language is a paradigm case of a biological adaptation, showing in its nature both its selective values and the ways in which it evolved, not necessarily by conscious design but piecemeal as the opportunities presented themselves. The breakthrough ideas came from the linguist Noam Chomsky (1957, 1966), some forty years ago, when he argued that all languages share the same "deep structure" and show themselves thoroughly biologically based in the ways in which we learn them and use them. Notoriously, Chomsky has been no friend of strict Darwinism, arguing that language ability evolved almost in a saltatory way, in one stroke. Later, more orthodox Darwinians think they can link things more tightly to selection. There is still much debate about how precisely this can occur. Some think the brain evolves and language follows; others think that language itself makes the running. To use an analogy, some see the key moves in the hardware with the software following; and others reverse the causal order (Pinker 1991, 1994; Deacon 1997).

Evolution of the Brain and Mind

Before we can resume our science/religion comparison, there is one final empirical question to be raised. This is a question suggested by the computer analogy. No evolutionist questions that the explosion of the brain in size was essentially adaptive. For all the costs, hominids with bigger brains had an adaptive edge over hominids with smaller brains. Precisely how the brain works has always been a matter of some debate; but now, in the computer age, there are many fruitful hypotheses showing how the brain can operate as a calculator to process and use information. One particularly popular thesis invokes the idea of the brain being built on the modular pattern. There is not one central unit doing everything at once, in an all-purpose fashion, but rather units which perform different tasks, and which are connected up in various ways (Fodor 1983).

What of the ultimate question, namely that of consciousness? Darwinians take consciousness very seriously. Consciousness seems so large a part of what it is to be a human that it would be very improbable that natural selection had no role in its production and maintenance. Even if one agrees that consciousness is in some sense connected to or emergent from the brain – and how could one deny this? – consciousness must have some biological standing in its own right. In particular, one would expect that perhaps consciousness started to emerge in a primitive way as animals developed bigger and better brains to do things – no one, for instance, thinks that ants are significantly conscious – and then was picked up by selection in its own right and developed and refined, perhaps pulling brains along in its wake to provide the material underpinning.

But what is consciousness, and what function does it serve? Why should not an unconscious machine do everything that we can do? Is consciousness just froth sitting on top of the brain's electronics? Is it a powerless epiphenomenon, to use the language of the philosophers? Almost certainly not. One major function is that of serving as a filter and a guide and coordinator to all of the information thrown up by the brain: consciousness sees that the brain does not get clogged up with overload of material that it does not need and cannot use. "Information must be *routed.* Information that is always irrelevant to a kind of computation should be permanently sealed off from it. Information that is sometimes relevant and sometimes irrelevant should be accessible to a computation when it is relevant, insofar as that can be predicted in advance" (Pinker 1997, 138).

My own sense is that consciousness is an important factor (part cause, part effect) in some of the very important human abilities to be discussed in later chapters of this book. I have in mind especially our distinctive ways of interacting socially (where morality comes into play) and of making choices in the face of alternative possibilities (free will or choice). Consciousness gives us a power and flexibility not possessed by those who do not have it. None of this of course explains consciousness as such, the reason for and nature of "sentience," as we might call it. Why should a bunch of atoms have thinking ability? Why should I, even as I write now, be able to reflect on what I am doing, and why should you, even as you read now, be able to ponder my points, agreeing or disagreeing, with pleasure or with pain, deciding to refute me or deciding that I am just not worth the effort? No one, certainly not the Darwinian as such, seems to have any answer to this. It is not to say that consciousness in the sense of sentience is denied – far from it. "Saying that we have no scientific explanation of sentience is not the same as saying that sentience does not exist at all. I am as certain that I am sentient as I am certain of *anything*, and I bet you feel the same. Though I concede that my curiosity about sentience may never be satisfied, I refuse to believe that I am just confused when I think I am sentient at all!" (Pinker 1997, 148). The point is that there is no scientific answer.

Philosophers have provided answers. Some, like Plato and Descartes, have argued that consciousness is a different thing, a substance apart from physical things: thinking substance (*res cogitans*) rather than material or extended substance (*res extensa*), to use the Cartesian dichotomy. However, although there have been such "dualists" recently, most notably Karl Popper and his friend John Eccles (1977), most people (including most Darwinians) feel uncomfortable with this philosophy. It is hard indeed to see how separate substances, mind and body, can interact. Most prefer to think of mind and body as manifestations of the same substance: they are "monists," subscribing to the "identity theory." Here they follow Spinoza, who argued that consciousness is a manifestation in some way of material substance. This is not to say that it is just material substance as traditionally conceived – thinking is not red or hard or round – but that it is part and parcel of material substance, and nothing more is added.

Where one goes from here is the difficulty. Although, given our interest, perhaps here is the point where legitimately we can draw back. The point is that consciousness is real, whether or not it is a separate thing, and it is something that seems open to the forces of evolution. More we cannot and need not say. We have to take it as a given, as of course we (as Darwinians) take the physical world as a given. Wonderful, mysterious,

familiar, all of those things and more: the unexplained starting point of
our inquiry.

The Soul

Let us swing over now to the Christian position on humankind, focussing
on a fairly conservative position, for there if anywhere will come the big
challenges. We have a special place in God's heart and in His creation:
"Then the Lord God formed man of dust from the ground, and breathed
into his nostrils the breath of life; and man became a living being" (Gen.
2:7). More than this, humans are made in the image of God. This does not
mean we are made in God's physical image – one does find some discussion in Aquinas (*Summa Theologiae* 1a, 91, 3.3) about the way in which
being upright is being more Godlike than being on all fours – but rather
in His mental or intellectual image. For this reason, although animals are
living, they are not made in God's image as are we. Bound up with
thought and reason is the capacity to act freely. We, like God, have choice.
Thus we are spiritual and moral beings.

It is an essential part of Christian theology that it is this rational soul
which makes us distinctively human: *imago dei.* It is this which survives
death and which provides the hope of immortality. Not that the body is to
be considered bad or irrelevant. In fact, Jewish thought always gave the
body a significant role. From this influence it is part of Christian belief
that, in some sense, after death body and soul will be one (Green 1998).
The soul will be, as it is now, "embodied" – although Saint Paul makes it
clear that while we now have a physical body, then we will have a spiritual
body. "It is sown a physical body, it is raised a spiritual body" (1 Cor.
15:44). It is also an essential part of Christian theology that the soul as we
have it now is corrupt or fallen. We are tainted by "original sin," which (in
the traditional story) dates back to the original human couple, Adam and
Eve. For all that they were living in bliss in Eden, deliberately they
disobeyed God and thus brought death upon themselves and their descendants: a bondage from which we were only freed by the sacrifice of
Jesus on the Cross.

Monogenism

What do we say about this in the light of Darwinian evolution? Although
the current Pope thinks that evolutionary thought can be reconciled with

Christian thought, he warns that this cannot be at the expense of Catholic teaching. "The sciences of observation describe and measure the multiple manifestations of life with increasing precision and correlate them with the time line. The moment of transition to the spiritual cannot be the object of this kind of observation" (John Paul II 1997, 383). Rather, "the experience of metaphysical knowledge, of self-awareness and self-reflection, of moral conscience, freedom, or again, of aesthetic and religious experience, falls within the competence of philosophical analysis and reflection, while theology brings out its ultimate meaning according to the Creator's plans" (383).

This is all very well, but what are the restrictions which are going to be placed upon an evolutionist? Can one possibly be a Darwinian under these conditions? You have got to spell out the relationship between mind and soul, and (if you are a Darwinian) apparently show how the one part evolved and the other did not. Then there are questions about immortality: when we die will it be just the spiritual soul which survives, or will the mind and intelligence be carried along as well? If the latter, the Pope would seem to be wrong about our miraculous origins. Mind and intelligence seem part of the soul package, and yet they are naturally produced. If the former, then can we truly be said to be made in God's image? The most important things seem not to survive death.

There are more problems. Francisco Ayala (1967), a distinguished evolutionary geneticist (and at the time of this writing, a Dominican priest), points out that an essential component of Christian theology, confirmed by Pius XII in his encyclical *Humani Generis* (1950), is that humans are descended from a unique pair (monogenism). That part of the Adam and Eve story cannot be interpreted symbolically. Moreover, there are strong theological pressures to go along with this conservative reading, otherwise one has removed a major support of the doctrine of original sin, namely that we possess it because we are descended from the original sinning pair, through the causal medium of sexual intercourse and consequent reproduction. In the words of a translator of Aquinas: "Theologically we are bound to accept a monogenist view of origins, that is one original couple from whom the whole human race has sprung, and whom we can conveniently call Adam and Eve. Though this is not in the strictest sense a point of Catholic faith, Catholics cannot reject it without, as far as we can see, compromising the doctrine of original sin, which *is* of faith" (Aquinas 1963, xxiv).

As Ayala points out, the trouble is that this goes completely against our thinking about the nature of the evolutionary process. Successful species like humans do not pass through single-pair bottlenecks: there is certainly no evidence that this was true of *Homo sapiens,* a species which seems to have been well spread around the earth. "There is no known mechanism by which the human species might have arisen by a single step in one or two individuals only, from whom the rest of mankind would have descended" (Ayala 1967, 15). More recently, "the genetic evidence indicates that human populations never consisted of fewer than several thousand individuals" (Ayala 1998, 36). Of course, theologically you can insist that some pair did uniquely get immortal souls (miraculously), and there is an end to it. By fiat you can introduce all of the intelligence you like into these souls. All else is contingent irrelevance. There were members of *Homo sapiens* before this pair and around this pair, but these others were not humans in a theological sense: that is, beings with immortal souls. But this stipulation is not without its difficulties, tensions certainly. Darwinian biology suggests that intelligence (and, as we shall see, related freedom and moral awareness) would be possessed by the parent generation and the contemporary generation and those of the next generations not descended from the pair. So on what basis can we declare them not to have been made in God's image?

And then, what right have we to say that this original sinning pair parented all the human beings living today? There is a hypothesis that there was an original "Eve" from which we are all descended (rather more than 100,000 years ago), but even if this be well taken – and the jury is still out scientifically – there is no guarantee of an original Adam (Loewe and Scherer 1997). Or if there were such an Adam, there is no guarantee that he and Eve lived together and mated, producing all the offspring. Perhaps Eve had two mates. Or perhaps Adam lived earlier or later. No one is claiming on the basis of science that we are descended from this pair alone. I am descended from my great grandmother, as is my sister, but we are descended from seven other people as well. This is not the way of the literal Adam and Eve story of Genesis – a pair of sinners from whom we are all descended (Ayala 1998).

We could embrace what is called Traducianism. Instead of assuming that each soul is created anew by God (creationism, in this sense), we might suggest that souls in some fashion are transmitted by parents. They are passed on as are other characteristics, like skin colour and height. This

view, irrespective of evolution, has its attractions, since clearly not everyone has a similar thinking apparatus, and we do tend to resemble our
parents more than others. Theologically, Traducianism has had favour in
some Protestant circles. It explains neatly the transmission of original sin:
if the soul is passed on, then the sinful soul is passed on (Hodge 1872).

From our perspective also, as Darwinians, Traducianism has its attractions. We can assert that the soul of man was something which was
transmitted, not just from the first man, but up from the animals. The
soul, in other words, evolved along with everything else. This meshes
nicely with what in this chapter we have seen that the Darwinian would
have to say about the evolution of the mind. Unfortunately, this belief or
escape route runs afoul of another part of Christian theology – Catholic
theology, at the least – namely that animals, although alive, do not have
(rational) souls. They do not mirror the divine. As one eminent evolutionist, who was also a practicing Christian, put matters: "It seems inconceivable . . . that a soul could be evolved by natural selection, hence this view
is usually coupled with belief in emergent evolution or a Life Force,
which to biologists is inadmissible; and since the claim that animals have
souls is also rejected by orthodox Christians on theological grounds, the
whole idea may, I think, be dismissed" (Lack 1957, 89).

We seem to have reached an impasse. And perhaps this was only to be
expected. "Reductionism" is the philosophy or methodology where the
aim is to explain away everything in terms of molecules and the like and to
deny reality to all higher level entities like minds and souls and so forth.
Darwinism, the apotheosis of a materialistic theory, is bound to be thor
oughly reductionistic. Virtually by definition, therefore, a religion making
souls central is bound to clash with a theory like Darwinism (Peacocke
1986). The essence of Christianity is that it is nonreductionistic, for minds
and/or souls do have genuine existence. We hardly needed to dig into the
theology to find that there would be difficulties. These could have been
predicted from the first. "With man, then, we find ourselves in the presence of an ontological difference, an ontological leap, one could say"
(John Paul II 1997, 383).

Reductionism

Paradoxically, what many would take to be the biggest problems may be
those which hold the most hope for reconciliation. Let us start with the

charge of reductionism, and its specific meaning in the Darwinian context.

The core belief in reduction is that one is explaining one set of things in terms of another set, and that the thing doing the explaining is more basic or fundamental than the thing being explained (Ayala 1974; Ruse 1973, 1988b). For this reason, the thing explaining is generally taken to be more uniform or less varied than the thing being explained, and reduction typically involves the bigger being explained in terms of the smaller and more standard and/or repeatable. Because of these various aims, we should make a clear distinction. When we are talking about reducing things rather than theories (which raises rather different issues), reduction might be taken at the *ontological* level, where what we are trying to do is to explain the many in terms of the few – many different things in terms of one or a few basic types, or substances – cr reduction might be taken at the *methodological* level, where the bigger is explained in terms of the smaller. These are not entirely separate aims or practices or beliefs, and they are certainly not in opposition. Methodological reduction might (usually does) involve or at least presuppose ontological reduction, and ontological reduction often seems plausible precisely because we have given a successful methodological reduction.

The Darwinian account of human nature is thoroughly reductionistic, both ontologically and methodologically, especially when it comes to the distinctive issues of brains and minds and so forth. The aim is to bring human thought down to two substances, probably just one. At the same time, or rather as and because one is doing this, one is explaining in terms of ever-smaller units: genes, obviously, but also minute brain parts. The brain is being subdivided into modules, and then these arc probably broken up even further, and then the whole is reassembled into the functioning human being.

However, notice what is going on, both ontologically and methodologically, and what is being claimed and what is not being claimed. Suppose, ontologically, we are going all the way down to one substance. No one thereby denies the fact of sentience – just its meaning and explanation (Schwartz 1991; Drees 1998). Hence, for the one-substance person (the monist) the notion of the basic substance, call it "material substance" if you will, is being expanded from simply solid, coloured, cold, and so forth, to something which includes sentience. There is nothing very surprising about this, especially in an age which talks in terms of quantum effects and of the complementarity of the electron and of the convertibility of matter into energy and so on. The old Cartesian notion of material substance as simple extended matter is long gone, anyway. But do recognize that this is happening.

Suppose now that Darwinian brain science proves a triumph of methodological reductionism. No one is then saying that the brain is nothing but a bunch of particles. The methodological reductionist is highly sensitive to *order,* more so than just about anyone else. Think of the triumph of methodological reductionism of this century, molecular biology founded on the Watson-Crick model of the DNA molecule as a double helix. The whole point about the DNA molecule – the point at which Watson and Crick hinted at the end of their classic paper (1953) – is that you get every different kind of gene, all the information you want, not from different subunits, but from the same subunits ordered in the chain in different ways. That is what the genetic "code" is all about. The same is true of the brain. Order the molecules in one way and you get junk. Order the molecules in another way and you get William Shakespeare.

What is important about being a methodological reductionist, and here we see a meeting with ontological reductionism, is that there is denial that the order represents a new kind of thing or substance. The order exists, it is not unreal, but it is not a thing in the way that a molecule is a thing. To think otherwise is to get oneself into that way of thinking which gives existence to such very nonuseful entities as those already-mentioned life forces, like the *élans vitaux*, that were popular in philosophical circles earlier in this century. The assembled and functioning DNA molecule is not a new substance. It is smaller substances (or substance parts), ordered. Of course, we do not necessarily recognize this – that is what scientific inquiry is all about! – and the reductionist is not saying that it is wrong to talk about entities at the macro level. It is just that the reductionist is a skim-milk person. He or she does not want to keep adding to the ontological richness of the world. Now you see the Eiffel Tower from the left, now from the right. It is not two entities mysteriously joined, but one entity seen in different ways. It is likewise with the DNA molecule and its parts. And analogous comments apply to us humans. The very crux of the Darwinian explanation of the distinctiveness of humankind is that we are ordered, and thus can function in ways that are not possible for other animals. It is not that we have something different at the substance level, but rather that we are different because of the way that we are put together: by natural selection for adaptive ends.

Augustine and Aquinas

Going back to Christianity and to a Christian understanding of the soul, there are two options (Murphy 1998). One is a kind of Augustinian position, in turn Platonic, which sees the soul as something more or less distinct, in a substance sense, from the body. Augustine depends strongly on the reading of "breath," a physical thing, in his exegesis of Genesis: "God fashioned man out of the dust of the earth and gave him a soul. . . . This he did either by implanting in him, by breathing on him, a soul which he had already made, or rather by willing that the actual breath which he produced when he breathed on him should be the soul of the man. For to breathe is to produce a breath" (Augustine 416–426, 503–4).

Unfortunately, if you take this kind of position, then you run into the kinds of problems already discussed in this chapter. One can be a Darwinian and be a dualist. How easy it would be to be a Christian of this kind and still be a Darwinian is another matter. However, this is not the only position for the Christian, nor even today is it the official Catholic position. The official position rather is that of Saint Thomas Aquinas, and he was heavily influenced not by Plato but by Aristotle. And never more so, given the influence of *De Anima,* than in his discussion of the soul. For Aristotle/Aquinas, the human soul – identified with the intellectual faculty, which makes a human a living human being – is not a thing, in the sense of a material substance. It is rather much more a principle of ordering or what, in Aristotelian terms, is called the "form" (Frede 1992). It is something real and can act as a kind of cause – Aquinas speaks of "actuating" – but it is not a substance. "Any particular body that is alive, or even indeed a source of life, is so from being a body of such-and-such a kind. Now whatever is actually such, as distinct from not-such, has this from some principle which we call its actuating principle. Therefore a soul, as the primary principle of life, is not a body but that which actuates a body" (Aquinas 1970, 7; *Summa Theologiae* 1a, 75, 1). And again, bringing in mind and body:

The soul is the ultimate principle by which we conduct every one of life's activities; the soul is the ultimate motive factor behind nutrition, sensation and movement from place to place, and the same holds true of the act of understanding. So that this prime factor in intellectual activity, whether we call it mind or intellectual soul, is the formative principle of the body. (Aquinas 1970, 43; *Summa Theologiae* 1a, 76, 1)

All organisms have souls as such: this is what makes them living. Only humans have "intellectual souls." This is the image of God.

The Soul as a Darwinian Concept

Let us start to bring things together. I do not want to pretend that Stephen Pinker or any other modern-day Darwinian is simply an Aristotelian. But I would say that the kinds of things we have been discussing in the evolution of the brain sound good to a modern-day Aristotelian. The key notion is information or order or form or some such thing. This is shared by Darwinian and Aristotelian. "My understanding of the soul is that it is the almost infinitely complex, dynamic, information-bearing pattern, carried at any instant by the matter of my animated body and continuously developing throughout all the constituent changes of my bodily make-up during the course of my earthly life" (Polkinghorne 1994, 163).

All of this explains why, for the Aristotelian, and the Thomist, and the Darwinian, the soul has to be or needs to be "embodied" in some way. It is not just a substance like a lump of rock or a lump of flesh. It needs to be activating, forming, informing, driving, and every other thing, making a lump of clay into a real human – as is appreciated by Saint Paul in his talk of the spiritual body. And what I suggest is that if you think of the soul in this Thomistic fashion as something animating the body, and that the distinctive human aspect of this is intelligence linked – as we shall see later – with freedom of moral choice, then this is very much in line with what the Darwinian sees as having evolved through natural selection. I do not find this coincidence surprising. Aristotle was not just a philosopher, but a biologist – and a good one (Gotthelf and Lennox 1987). He was also someone whose philosophy was "naturalistic" in the sense that he tried to base his philosophy on the empirical facts as he saw them. His notion of soul was not intended to be airy-fairy or ethereal, but something which makes sense of things as we find them. This is the Darwinian position also. There are also the functional or teleological links which I shall discuss in a later chapter.

There is work to be done. One can see ready ways of exploring how the soul as now conceived might survive death – God is not keeping up a substance but rather information which, at some point, can be reactivated – but in other respects there are still potential tensions between Darwinism and Christianity. If the soul is now an empirical thing, in the fashion just discussed – empirical, but not just a material thing – then its miraculous origins as demanded by Catholicism seem even more distant and implausible. One has walked right into the claim that the soul evolved from lower organisms, by unbroken law. One goes smoothly from animal souls to human souls. Traducianism squared! Not that this evolution of the human soul cannot be a God-backed process. Nor that logically one cannot arbitrarily start human souls off miraculously at some specific point in time. So I do stress that we have tensions rather than absolute and ineradicable contradictions. But, having bought into Darwinism thus far, the inclination is to think that there has been a gradual upward development from organisms with less sophisticated principles of ordering and thinking. The miracles are not one-of-a-kind events, but part of everyday life.

There will be more in the next chapter about miracles, and more in an even later chapter about original sin – one of the points that is a stumbling block to any evolutionary account of the human soul. So for now, let me conclude this part of the discussion. There is that in Darwinism which should attract the most conservative of Catholic thinkers on the human or intellectual soul. Our understanding of the mind/brain is going ahead at such a speed that it would be foolish today to make any definitive and final judgements, scientific or theological, on these questions. In the spirit of Augustine on interpretation and of Aquinas on metaphysics, it behoves Christians to see just how rigid are those theological beliefs which seem to go against a full-blooded Darwinism. Speaking positively, one should keep an open mind to the possibility of making the reductionism of Darwinism a point of Christian strength, rather than regarding it as something to be avoided as a vampire avoids garlic.

Contingency?

We are not yet finished with humans. There is another matter which demands our attention. Although the Christian can agree that we humans are animals, we are not just any old kind of animal. We are animals of a very special kind. We are the focus of God's love and care and attention. We are the beings for whom He suffered on the Cross. All of this means that for the Christian we humans are not contingent beings. It is not by chance that the universe exists, and it is not by chance that we exist within the universe. We are the focus and purpose of creation. God cares about us, and it would be unthinkable within the Christian scheme were humans not to exist. "However many billions of galaxies there are, they will all exist as parts of a process whose goal is the existence of human beings" (Ward 1998, 22). It is not necessary that we actually be the *Homo sapiens* that we are today. God's purpose would surely be possible and satisfied if we had only three fingers and a thumb on each upper limb. But we must be humanlike. If humans or their alternatives were not intelligent, God's design would not be satisfied. Totally nonintelligent beings would lack the freedom so crucial within the Christian system, quite apart from being unable to do such things as worship the Creator. Humans or humanlike creatures within certain limits must exist.

But is not evolutionary theory, Darwinian evolutionary theory in particular, in some sense – in some very deep and radical sense – contingent? Surely Darwinian evolution is nondirected or nonprogressive in such a way that there is simply no guarantee that humans or anything else would have evolved. Indeed, is not the contingency so radical that it is highly improbable that humanlike creatures would have evolved? Darwinism is evolution through natural selection working on random mutations. Non directionality comes first from the randomness of mutations. They are not random in the sense of being uncaused, but they are random in the sense of not appearing on demand according to need. The vast majority of mutations are deleterious; only rarely do they help their possessors.

Nondirectionality comes second from natural selection. It is a relativistic, opportunistic mechanism. What succeeds is what succeeds. Remember the finches. If there is a drought, then big strong beaks count. If there is food aplenty, then something else counts. There is no ultimate absolute better or worse, high or low. Nondirectionality comes third and finally from external circumstances, be these other organisms or forces of nature. The dinosaurs were doing very nicely, and the mammals were continuing their way of 150 million years of undistinguished nocturnal ratlike existence, when an asteroid fell on earth and altered things dramatically. Analogously, life was good for the giant birds of New Zealand until the Polynesians arrived on their doorstep.

On the Darwinian picture, there is simply no guarantee that human or humanlike creatures – creatures with intelligence and a moral sense and so forth – would evolve. Indeed, the chances seem slim indeed. For all that he claims that there is no conflict between science and religion, Stephen Jay Gould puts things in an uncomfortably blunt fashion:

Since dinosaurs were not moving toward markedly larger brains, and since such a prospect may lie outside the capabilities of reptilian design (Jerison, 1973; Hopson, 1977), we must assume that consciousness would not have evolved on our planet if a cosmic catastrophe had not claimed the dinosaurs as victims. In an entirely literal sense, we owe our existence, as large and reasoning mammals, to our lucky stars. (Gould 1989, 318)

Tentative Solutions

How are we going to solve this problem? A number of solutions offer themselves. The first and second work from empirical premises; the third and fourth are more theological, bringing in God's action. Which direction you will prefer, empirical or theological, will depend obviously in part on how satisfactory you judge the various solutions to be, but also in part on your general views with respect to the science/religion interface (Barbour 1988). If (like Dawkins) you believe that science and religion can overlap in interests and solutions, then you will take seriously the empirical approach. Science and religion must at a minimum be brought into harmony, if not indeed mutual support. You are going to show that Gould, qua science, is just plain wrong about chance and "lucky stars." If (like Gould) you think the two, science and religion, must be kept apart, then you will probably incline to a theological solution. In some way, you are going to show that Gould's chance is a scientific chance and does not imply a theological chance. Our lucky stars are no luck in the eyes of God.

No doubt there will be feedback here. If you think the empirical solution works better, you will be more favourable to a science/religion connection view. If you think the theological solution works better, you will be more favourable to a science/religion independence view. Neither attitude, as such, dictates whether a Darwinian can be a Christian, and of course whichever position you take will have some influence on the solution you favour. Just keep in mind that your specific judgement on what Darwin says about the status of humankind is part of a more general attitude toward the science/religion relationship, even if your attitude is (as I suspect is the case for many of us) somewhat ecumenical. Sometimes the overlap position looks right, sometimes the independence position looks right!

On the empirical front, first, we might suppose that evolution, even Darwinian evolution, is a lot more directed than someone like Gould allows. Many evolutionists in the past have argued that it is directed upwards towards humans, usually invoking special directing forces of a divine or quasi-divine nature (Bowler 1983). This is no longer acceptable, nor is there anyone in the scientific community who thinks such forces necessary. For all the inspiration that some very distinguished members of the evolutionary community have drawn from his approach, the French Jesuit paleontologist Pierre Teilhard de Chardin (1955) was criticized without mercy when he tried to show that evolution moves up from the biosphere, through the noösphere (human culture), to something he termed the "Omega point," which he identified with Jesus Christ (Medawar 1961; Ruse 1996a).

But, from the very earliest days, there have been many who have tried to get nonguided direction for evolution. Beginning with Darwin himself, there have been those who have tried to get such direction through the causal medium of natural selection (Ruse 1996a). Today's most ardent proponent of such selection-driven directionalism is Richard Dawkins – an interesting irony, given the views he holds about the Christian religion. He is an enthusiast for biological "arms races," believing that evolving lines of organisms compete against each other, improving adaptations: as the prey gets faster, so also the predator gets faster; as the shell gets thicker, so also the teeth and jaws get stronger. Overall, it is thought that this kind of comparative progress leads to a kind of absolute progress, which is to be expressed in terms of brains and intelligence (Ruse 1993). Dawkins especially draws attention to the way in which military arms races have evolved, from focussing on such things as more efficient armour and weapons of destruction, to the use of computers and other electronic hardware (Dawkins and Krebs 1979; Dawkins 1986, 1997a). So likewise, in the animal world, we have the evolution of organisms with ever greater and more powerful on-board computers. Humans or humanlike creatures may not have been absolutely necessary, end consequences of evolution – presumably somebody might have won all of the arms races earlier on – but their appearance is far from a matter of brute coincidence. They are just the sorts of things one might reasonably have expected.

Backing this line of approach, many Darwinian evolutionists argue also that natural selection does not mean that organisms can evolve in every possible direction. They are "constrained" by the need to find the right adaptive niche, and not all such niches are equal. This is why one finds so many cases of "convergence," where organisms of vastly different origins take what are essentially the same adaptive routes: the fish and the marine mammals most famously, and the placental mammals and the marsupials most dramatically. This is not simply a question of something like a giant bird (like a moa) taking the ecological niche of a mammalian predator (like a wolf), but of real physical similarity: the sabre-toothed tiger and the marsupial "tiger" of South America, for instance. "Although there may be a billion potential pathways for evolution to follow from the Cambrian explosion, in fact the real range of possibilities and hence the expected end results appear to be much more restricted" (Conway Morris 1998, 202). Consequently, "within certain limits the outcome of evolutionary processes might be rather predictable." Dare one say that we alone have evolved to occupy the humanlike creatures niche?

The other empirical option is to appeal to the vastness of the universe and to argue that, although someone like Gould may be right on a global level, at a universe level – as Gould himself seems to think – his worries are much less troublesome. "I can present a good argument from 'evolutionary theory' against the repetition of anything like a human body elsewhere; I cannot extend it to the general proposition that intelligence in some form might pervade the universe" (Gould quoted by Dick 1994, 395). Surely, on some of the billions and billions of planets that must exist in the universe, life must have evolved as it did on our Earth. On this world of ours, life must have started just as soon as it possibly could. If it started so soon down here, then surely it must have started occasionally elsewhere. And this being so, who dare say that consciousness has never or could never have appeared elsewhere? Hence, who dare say that it was pure chance or contingency that we conscious beings exist?

Moving now to the more theological options, one suggestion – endorsed particularly by a number of theologians with backgrounds in physics – is that somehow God's design is built into the process at the quantum level. This is a position which stands in line back to Darwin's great American supporter, the Harvard botanist Asa Gray (1860). He argued that the raw material on which selection works, what we today would call "mutations," is directed by God. Selection, as it were, weeds out the unneeded and inadequate. Today's thinkers do not want to put down change directly to God's intervention, in the sense of violating or otherwise disturbing the regular course of nature. However, they argue that quantum indeterminancy leaves open a space for God to act in a positive fashion. The mutations that are needed are caused by quantum-level effects – it is just that, overall, they do not stand out because they are averaged by non-needed or nondirected changes. Suppose one needed a positive change at time t. Overall, quantum theory might tell you that you will get five positive changes and five negative changes, but that we cannot predict which will be positive or negative on any particular occasion. There is nothing to stop God from putting in the positive change when He wants, namely at time t, so long as He lets it be masked by the other nine positive and negative changes. Laws are not being violated and special directed laws are not needed, but direction is coming nevertheless (Russell 1998; also Polkinghorne 1989).

[handwritten margin note: isn't the neg "naused" by God too? What "mask"]

Finally, one might propose a more Augustinian solution. When God created, He foresaw all that would happen and was quite capable of intending humans to appear, even though the normal processes of evolution would not guarantee such an appearance. The fact that we have appeared shows that it was possible, and this is all that was needed to make it actual. God Himself is outside time, and so in the act of Creation could will human appearance, for He is involved in the world immanently through all of our time.

It makes no difference, therefore, whether the appearance of *Homo sapiens* is the inevitable result of a steady process of complexification stretching over billions of years, or whether on the contrary it comes about through a series of coincidences that would have made it entirely unpredictable from the (causal) human standpoint. Either way, the outcome is of God's making, and from the biblical standpoint could properly be said to be part of God's plan. (McMullin 1996, 156–57)

God is not simply forecasting on the basis of what will happen. There is an act of creation which unfurls through time for us, but which is outside time for God and hence for which beginning, middle, and end are all as one. "Terms like 'plan' obviously shift meaning when the element of time is absent. For God to plan is for the outcome to occur. There is no interval between the decision and completion. Thus the character of the process

which, from *our* perspective, separates initiation and accomplishment is of no relevance to whether or not a plan or purpose on the part of the Creator is involved" (157). Hence "the contingency or otherwise of the evolutionary sequence does not bear on whether the created universe embodies purpose or not. Asserting the reality of cosmic purpose in this context takes for granted that the universe depends for its existence on an omniscient Creator."

Darwinian Direction

Let us look at these suggestions, starting at the empirical end of the scale. Arms races and many worlds are complements and not contradictories. They are both trying to solve the problem from the side of what seems scientifically reasonable. I will put aside the many-worlds argument at this point, for it will be the subject of a later chapter. What about arms races? There are actually two issues here. The first is whether it is reasonable for the Darwinian to believe in any kind of biological progress whatsoever. Is it acceptable for a selectionist to think that nature's path has been from monad to man? Then second there is the question of whether (appearances to the contrary) the Darwinian can claim that such progress comes about through selective mechanisms. Specifically, will arms races do the job – or arms races backed by constraints on possible paths, searching out appropriate niches?

Evolutionists, including Darwinian evolutionists, are badly split on the question of whether or not the path of evolution is progressive, from simple to complex, from the blob to the human. Gould (1988, 319) is scathing. He speaks of progress as "a noxious, culturally embedded, untestable, nonoperational, intractable idea that must be replaced if we wish to understand the patterns of history." It is a delusion engendered by our refusal to accept our insignificance when faced with the immensity of time (Gould 1996). Edward O. Wilson, at least as distinguished a Darwinian, takes completely the other tack:

The overall average across the history of life has moved from the simple and few to the more complex and numerous. During the past billion years, animals as a whole evolved upward in body size, feeding and defensive techniques, brain and behavioral complexity, social organization, and precision of environmental control – in each case farther from the nonliving state than their simpler antecedents did. (Wilson 1992, 187)

He concludes: "Progress, then, is a property of the evolution of life as a whole by almost any conceivable intuitive standard, including the acquisition of goals and intentions in the behavior of animals."

Wading into treacherous waters, my opinion is that if evolutionists want to argue for progress, then really nothing can stop them. This is not necessarily to say that one is going to accept any particular measures of such progress, other than a vague intuition that progress means becoming more humanlike. But inasmuch as the notion of progress is read into the science rather than derived from it, there can be no true objection. However, there is one cautionary point which should be mentioned. There is no logical connection between a belief in biological progress and the broader cultural ideology of progress, the belief that society is improving through human effort. Indeed, Gould is against the idea of biological progress precisely because he believes it has built-in racist overtones inimical to social progress! But, historically, biological and social progress have come together, and there is no doubt that today's most enthusiastic biological progressionists (Wilson notably) are also social progressionists, seeing the two as part of the same world picture (Ruse 1996a).

Unfortunately, traditionally, social progress has been thought incompatible with Christian theology, which stresses rather God's providential actions in the world (Bury 1920). For the progressionist, improvement through unaided human action is possible, whereas for the Christian this is Pelagianism, contradicting the belief that only through God's undeserved grace is any true improvement possible. So I do caution. You might be happy accepting biological progress but denying social progress, or (as was the wont of many late nineteenth-century liberal Christians) you might claim that social progress is not really incompatible with true Christian faith; but you should be aware that there are shoals here posing dangers for those ardent to reconcile Darwinism and Christianity.

Let us turn now to Darwinian mechanisms for biological progress, specifically arms races. Scientific opinion seems to be divided on the subject (Ruse 1993). The invertebrate paleontologist Geerat J. Vermeij (1987) argues that the fossil record shows clear evidence of "escalation," where new and improved characters are developed by competition between lines. Overall, "if selection among individuals predominates over other processes of evolutionary change, and if enemies are the most important agencies of this selection, the incidence and expression of traits that enable individual organisms to cope with their enemies (competitors and predators) should be found to increase within specified habitats over the course of time." He adds, "On the whole, the evidence from fossils is in accord with the hypothesis of escalation" (359). Vermeij details improvement in metabolic rate, shell-breaking capacity, shell-repair ability, dental specialization, locomotive ability, and other features.

However, even if one has this kind of comparative progress, as one might call it, nothing guarantees that eventually it will all add up to absolute progress, meaning the evolution of brains and so forth: the evolution of humanlike attributes. Breaking shells and doing philosophy are worlds apart. Moreover, arms races have their critics. Steven Stanley (1979) calculates that the fossil record does not show that the classic predator-prey arms race – carnivores versus herbivores – manifests any real improvement over millions of years. Paleontological studies do not suggest that anyone got that much faster after things had got under way. Others, like Douglas Futuyma and Montgomery Slatkin (1983), argue that ecological conditions are always so intertwined that any expectation that one line will compete in isolation, as it were, against another line is simply doomed to disappointment. Certainly this cannot happen over the long period surely required before we might expect genuine improvement.

Sharing the somewhat ambiguous status of arms races is the subargument about constraints and channelled paths of evolution. This is an area where many different pieces of information are starting to come together, especially about the nature of the underlying molecular genetic basis of organisms and how this cashes out in development. It is becoming increasingly apparent that putting together a functioning organism means using parts – particularly genetic parts – which occur right across the range of living kinds. Famously (or notoriously), virtually the same genes code for eyes in house flies and mice (Carroll 1995). So if once you have got the basic organisms up and running, perhaps there is already the potential for a humanlike being. And if you combine this with the fact that we do now have measures which show that organisms do not occupy all of the possible places in morphological space, but rather that they tend to cluster together at certain "hot spots," as one might say (Foote 1990), then perhaps the way is paved for evolution always going in certain set directions, no matter what the incidental contingencies.

So, what should we conclude on the scientific side? As a Darwinian, you can build a picture that human emergence was more than chance. Arms races are probing upwards, and a human-organism-type niche – perhaps culture, akin to sea, land, and air – lies waiting for the arrivals. And the delicious irony is that this is science endorsed enthusiastically by scientists (Dawkins in particular) who would not welcome the implications drawn by the Christian. But it is all a bit "iffy." Even if arms races do work, there is a large gap between comparative and absolute progress. One would also like to know why constraints would direct one upwards towards intelligence. Internal constraints (the somewhat contingent effects of the very mechanism of getting development to occur) seem no more directed upwards than downwards. And external constraints (the limited availability of functioning niches) raise the question of the assumption that culture is an ecological niche of the same order as sea, land, and air. Indeed, some evolutionists question the very notion of an empty niche, waiting to be occupied, arguing rather that organisms themselves create the niches.

A modest but positive conclusion is indicated. The Christian would be foolish to think that Darwinism insists that humans are uniquely significant and bound to appear. However, the Christian can find in Darwinism some support for the belief about the special significance of humans and the probability of their appearance. And for those Christians who share with me a sense of unease that so crucial a part of their world picture should be supported by so unchristian sounding a mechanism as an arms race, may I point out that this talk is metaphorical – no one is saying that conscious (perhaps bad) intention is involved in the development of biological weaponry (or "weaponry") – and that while violence is obviously involved (as when predator and prey interact), this general point will get full discussion when we turn to such issues as pain and evil.

The Augustinian Option

Turning now to the options which include theological suppositions, I will dismiss at once the suggestion appealing to modern physics: that which supposes that God might slip in a directed quantum event as it suits His purpose. This is very much a "God of the gaps" kind of argument: if you cannot think of an explanation of how things work, then let us see if you can fit God into the spaces where your understanding fails. But as so often with simple solutions, there are complex difficulties. Apart from anything else, if God can slip in to produce the right quantum event when a mutation pointing to humankind is needed, why can He not likewise slip in to prevent the wrong quantum event when a mutation causing great pain and unhappiness is about to occur? God gets credit for producing intelligence. Why should God not also get blamed for producing sickle-cell anaemia?

The Augustinian option is more sophisticated. It says that, no matter why things are as they are, the fact is that humans evolved and did so by force of law: no doubt, in conjunction with all sorts of random factors. Whether one wants to put down the success to a one-of-a-kind triumph over randomness or whether one thinks that the evolutionary process is such that randomness can be overcome, the fact is that human evolution did happen. And this was God's work. He is not an impersonal God, planning all beforehand and then letting it happen, but He is actively involved in seeing that things occur as they should. The laws and the events, random or not, are His laws and events, and He foresaw and intended the end result. He therefore deserves credit. The point here is that it really does not matter what the science may turn out to be, God is going to win!

Here also, however, there are costs to the solution, costs which must be paid by any Augustinian. In the first place, the whole question of freedom – especially the freedom of the human will – is thrown into doubt. Does God control everything to such an extent that things had to happen by the divine will, no matter what anything within His creation wanted or intended? Or does one want to say that free will came into play only when humans had finally appeared? In the second place, there is again the question of evil and pain. Even if we put much evil down to the actions of free humans – Hitler and Auschwitz and that sort of connection – there still remains natural evil or pain. If God so controls things that humans appeared despite the improbability, then is God also responsible for such things as bad mutations? Is God responsible for those earthquakes and floods and other natural calamities which take so many innocent lives? Making God responsible at one level seems to entail making Him responsible at other levels also. Can one truly say that the only way in which God could have made humans was by some process which involved individual mal-mutations? Is this compatible with God being a Father, as is claimed by Christians?

These problems are not new problems, but they are problems that in some fashion must be answered by the Christian. However, rather than rushing into an immediate and full discussion, I ask you to wait until later chapters. In due time they will get full attention. At this point, we have said all that can be said directly on the question of humans and their evolution. Hence, the discussion of this chapter can come to an end. We have seen no absolute bars to the Darwinian being a Christian, and we have seen some (perhaps surprising) points where Darwinism and Christianity, even conservative Christianity, resonate. But, even apart from a number of yet-to-be-discussed questions, anyone who thinks that matters are easy and settled has simply not been paying attention to my arguments.

CHAPTER FIVE

Naturalism

We move on now to the central Christian drama. God, seeing us in a state of sin, became incarnate in the human form of Jesus Christ, lived and preached and then was crucified for our benefit, rising again on the third day. Darwinian evolutionary theory is simply irrelevant to much of this story. How we should interpret God's death, for instance: as a sacrifice, as a substitute, as a ransom, or what? But Darwinism does impinge on the story in very important respects. Most obviously, there is the problem of miracles. The Christian story tells of Jesus born of a virgin, turning water into wine and feeding the five thousand and raising the dead (and of some of his disciples being able to do some of these things also), and most significantly coming back from the dead himself just a short while after he had been taken down from the Cross and buried. Darwinism is a theory committed to the ubiquity of law. In the language of the philosophers, it is a "naturalistic" theory. How can it be reconciled with a world picture so obviously committed to the breaking of law?

Miracles

As always in philosophical discussions, a lot depends on the meaning of terms. By "law" in this context we mean scientific law, and this means a universal statement referring to a regularity of the empirical world, which in some sense is both true and necessary (Nagel 1961; Hempel 1966). Although mathematics is used by scientists, it does not itself contain scientific laws because it is a system of pure ideas. A true statement of fact, like "the Eiffel tower stands in Paris," is not a law, because it is not a regularity. And a true regularity, like "all of my children are under thirty," is not a law, because although it is true and a universal claim about the empirical world, it is not a necessary statement. At some point, it will be false. Newton's law of gravitational attraction, about all bodies feeling an attractive force inversely proportional to the square of the distance between them, is a scientific law. If we have an apparent exception like a boomerang not falling at once to the earth, we look for intervening factors, because without such factors we know that the object must fall. And Darwinian theory is certainly intended to be a body or network of such laws. Natural selection is something which is going to hold when you have populations of organisms, and Mendelian/molecular genetics does not fluctuate as the seasons wax and wane (Ruse 1973, 1975a).

Miracles are generally taken to be in some sense violations of or exceptions to law, brought on by divine desire and agency (Swinburne 1970). But notice the key qualifiers "generally" and "in some sense." If Lazarus, whom Jesus raised from the dead, was stinking rotten with maggots crawling through him and then got up and walked, this is a straightforward use of miracle, no question. The laws of nature have been broken. Likewise if Jesus really did walk on water, we have such a miracle. But not all miracles, including not all Christian miracles, have been of this nature. Remember the Catholic doctrine of transubstantiation. The belief is that the bread and the wine of the communion service turn, literally, into the body and blood of Christ, the "host." In Aristotelian terms it is the essence, the substance, which changes; the accidents, the contingent attributes remain untouched. The host remains breadlike and winelike, all the way through to the naked eye and even under the most intense microscopic investigation. Something is happening outside of the laws of nature, perhaps, but no laws as we know them are being broken.

And many would want to extend the notion of miracle even to certain events or phenomena where everything happens in accordance with law. When I was a child, many people thought that Dunkirk – when the British army escaped in 1940 from France, in the face of the all-conquering Germans – was a miracle. The English Channel, often so very rough, was exceptionally calm, enabling the most fragile of craft to cross and to bring back servicemen. If someone had pointed out that one could give an explanation in terms of natural laws, I do not think that these people would in any way have changed their firm conviction that the Channel that weekend was miraculously calm, in the sense that God acted to preserve the army to fight again against the evil Nazi hordes. They would think that that only substantiates God's power and magnificence and love of His creatures. What counted was not whether or not laws were being obeyed or violated, but the context: the meaning. This was a time when England stood alone, and had God not taken personal responsibility for events at that moment, then the whole world might have been on the way to despair and destruction.

Agreeing fully that Darwinism pushes the rule of nature's laws, where does Christianity stand on this? How does a believer handle the question of miracles? Many post-Enlightenment believers, and this would include today's more liberal Christian theologians, would be strongly inclined to the miracle-compatible-with-law stance just sketched. Apart from the fact that one thereby avoids all of the problems about new souls, they would argue that the many miraculous events in the Gospels do not require nor are they improved by being miracles of the violation-of-law variety. Many if not all of the miracles happened according to law; their miraculous nature comes from their meaning or significance. The everyday miracles of the Gospels – turning the water into wine and feeding the five thousand and even raising Lazarus – can be explained as the enthusiasm of the moment. People's hearts were so filled with love by Jesus' talk and presence that spontaneously and out of character they shared their food. To think otherwise – to think that Jesus actually turned loaves and fishes into a banquet – is if anything a bit degrading, making the Redeemer a kind of high-class caterer. Lazarus and the ruler's daughter were more than likely brought back from trances. They may have been dead to all intents and purposes, and Jesus' actions were highly significant, but one should not suppose that Lazarus and the girl were necessarily clinically dead.

In fact, even the supreme miracle of the resurrection requires no lawbreaking return from the dead. One can think Jesus in a trance, or more likely that he really was physically dead but that on and from the third day a group of people, hitherto downcast, were filled with great joy and hope. That a psychologist or sociologist might be able to explain all of this by natural laws is totally irrelevant – something of a relief, actually. What counts is that it happened and that it was unexpected and that it mattered. Conjuring tricks are beside the point. It is from this regeneration of spirit that true Christianity stems, not from some law-defying physiological reversals in the early hours of a Sunday morning (O'Collins 1993).

This is not a position of desperation. It is one which finds real positive value in such an interpretation of miracle. And of course part of the value is that one is meshing with science. Darwinism as such has little relevance to this issue, but its underlying naturalistic philosophy is a key factor in the theology of the liberal thinker whose position has just been sketched. But what if one cannot go this far? What if one belongs to one of the more conservative categories in the taxonomy of Christians sketched in an earlier chapter. What if one feels that many (most, all) of the miracles did involve a breaking of natural law? One feels that they simply cannot be explained away as natural, or that they stand outside of (beyond but not violating) the law. If the virgin birth of Jesus were natural, presumably it would start with the spontaneous dividing of an ovum of Mary, which would mean that the child would be an XX chromosome bearer and hence female. One feels, therefore, that the birth of Jesus had to be more than natural or non-lawbreaking. But in any case, for such a person there is a strong theological objection to explaining away the miracles as natural. The whole point is that Jesus was well and truly dead and then rose miraculously back to life. His conquering of death was precisely that which makes sense of the whole story of the Incarnation and Atonement. (This is the position of the important German theologian Wolfhart Pannenberg [1968].)

Even here, I simply do not see why one should having any trouble accepting that the law-governed world, including the evolutionarily law-governed world, is the background against which the miracles of Christianity occurred. Note that the main objection (for liberal believers) to the miracles being miraculous (sensu violations of law) seems to be the fact not that they are miraculous and hence antiscience, but that they are theologically offensive. They involve God being a conjurer or some such thing. If this is not a worry, then Darwinism hardly exacerbates the problem. It is true that one's acceptance of the miracles of Christianity (sensu violations of law) is not exactly something based on foolproof argument; but, after people in the eighteenth century made such devastating attacks on law-breaking miracles, one knew that all along. There are quite enough problems with the authenticity of the miracles because of the ambiguous (perhaps corrupt) nature of the biblical texts, making them dubious as a matter of pure reason, without the added difficulties of science and its commitment to law. If you are prepared to accept the miracles (as law-breaking miracles) despite these difficulties, I doubt that science is going to make much difference anyway.

Indeed, as for the liberal thinker, there are strands of theological thought which rather reinforce one's position, welcoming the scientific background against which miracles (*sensu* law violations) supposedly occur. In the first place, one might say that the whole point of miracles is that they are miraculous. If they are occurring all of the time, then the miracles of Jesus are hardly that exciting or significant. It is precisely because they do not occur as a matter of course – that the world is so law-bound – that they become particularly significant. In the second place, complementing the first point, one can make a traditional distinction between the order of nature and the order of grace. That is, between what is known as cosmic history and what is known as salvation history. "The train of events linking Abraham to Christ is not to be considered an analogue for God's relationship to creation generally. The Incarnation and what led up to it were unique in their manifestation of God's creative power and a loving concern for the created universe." Because and precisely because we as free beings had sinned, a special intervening act was required of God. "Dealing with the human predicament 'naturally', so to speak, would not have been sufficient on God's part" (McMullin 1993, 324). It goes without saying that the creation of animals and plants was an entirely different matter and that there was no call here for miraculous intervention.

Atheism

Not everyone will be happy with this synthesis or attempt at harmony. There are both Darwinians and Christians who argue that if one starts using law, becoming a naturalist, this is the slippery slope which ends at the bottom with materialism: meaning at this point that nothing supernatural at all exists, which means atheism, which means that Christianity is ruled out as false. Hence, Darwinism, as a supreme manifestation of the naturalistic philosophy, ends in the falsity of Christianity.

On the one side, we find the historian of evolutionary biology William B. Provine. He does not hesitate to label as "intellectually dishonest" those who think that law and God are compatible. His is a kind of theology by stipulative fiat: "A widespread theological view now exists saying that God started off the world, props it up and works through laws of nature, very subtly, so subtly that its action is undetectable. But that kind of God is effectively no different to my mind than atheism" (Provine

1988, 70). Having made this judgement, Provine has little trouble finishing the case, concluding that this new God "does nothing outside of the laws of nature, gives us no immortality, no foundation for morals, or any of the things that we want from a God and from religion." The honest Darwinian is an atheist, in practice if not in name.

On the other side, we find Phillip Johnson. Speaking of naturalists, he notes that they "concede that some problems are not yet solved, but they are confident that science will solve them by proposing natural mechanisms because science has so often been successful in the past. Bringing God or intelligent design into the picture is giving up on science by turning to religion (miracle) and invoking a 'God of the gaps.' The Creator belongs to the realm of religion, not scientific investigation" (Johnson 1995, 208). From here it is but an easy step to atheism.

Naturalism is a *metaphysical* doctrine, which means simply that it states a particular view of what is ultimately real and unreal. According to naturalism, what is ultimately real is nature, which consists of the fundamental particles that make up what we call matter and energy, together with the natural laws that govern how those particles behave. Nature itself is ultimately all there is, at least as far as we are concerned. To put it another way, nature is a permanently closed system of material causes and effects that can never be influenced by anything outside of itself – by God, for example. To speak of something as "supernatural" is therefore to imply that it is imaginary, and belief in powerful imaginary entities is known as superstition. (37–38)

What is Darwinism? It is atheism!

Surely there is a response here. As a scientist, as a Darwinian, one is committed to the rule of empirical law. But this is aside from whether one thinks that there is a reality beyond this law. As a Christian, one thinks that there is more. Yet, even if one wants to argue for rule-breaking miracles imposed by grace on top of the order of nature, one is trying deliberately to keep these beliefs separate. Let us therefore speak of "methodological naturalism" and of "metaphysical naturalism." The metaphysical naturalist is the person who is an atheist, who does deny that there is anything beyond blind law working on inert matter. The methodological naturalist, who may well be an ardent Darwinian, is one who states that for the purpose of doing science nothing but law will be entertained, but who recognizes that there might be more, in fact or meaning.

Johnson has little time for this move. It is true that "no one can disprove that sort of possibility"; unfortunately, "not many people seem to regard it as intellectually impressive either." Faith is pushed to the side, confined to private life and out of the public gaze. But in response, our question must be whether all attempts at reconciliation are quite as inadequate as Johnson, and Provine coming from the other side, suggest. Ignore the breathtaking theological arrogance (or ignorance) which brazenly declares that such moves – surely in the spirit of Karl Barth, the greatest theologian of the twentieth century – are not regarded as "intellectually impressive." Simply making such a disparaging claim may be rhetorically satisfying, but it is hardly philosophically adequate. More argument is needed before we must accept the shared negativism of our two stern critics. And expectedly, when we turn to the philosophers, we find that precisely such argument has been offered. Most vocal has been Alvin Plantinga (1997), and as always he must be taken seriously. He claims that even if you do not think that methodological naturalism necessarily leads to denial of God – pitchforks you into metaphysical naturalism – at a minimum there is a tendency that way, and at a maximum there are theoretical and pragmatic reasons why you will go that way. Hence, no matter what one might think, qua naturalism Darwinism is no friend of Christianity. Let us see how he makes this case.

Augustinian Science

Plantinga has a number of arguments, and it is important to understand their overall intent. He is not – at least, he claims he is not – against science as such. His objection is to science which makes central the rule of law. His objection is to naturalistic science, meaning at the least what we are now calling methodologically naturalistic science. Darwinism is highlighted as a paradigmatic example of such a science. Plantinga himself would substitute something which he calls "Augustinian science," which allows for miracles as well as laws – specifically, which allows for Christian miracles – and which apparently is rightfully so labelled because this is the science which Saint Augustine would endorse were he alive today. Because Augustinian science is not even methodologically naturalistic, there is no temptation, much less compulsion, to slip over into God denial, what we are calling metaphysical naturalism.

What is the objection to Augustinian science? One major complaint made by Plantinga is that Augustinian science is simply ruled out by definition: in a quite arbitrary fashion, something like Darwinism unfairly delimits the boundaries of real science. By fiat, science is characterized as something produced according to a methodologically naturalistic philosophy. Christianity is therefore put beyond the bounds of science. Even if it is not declared logically impossible, it is made into second-class knowledge. It is in some sense belittled. As a Darwinian, I myself am named as a major culprit in this respect, and my Arkansas testimony is highlighted as particularly offensive. "Science" is defined as "miracle excluding," and then it is triumphantly announced that Christianity is not science!

One thinks this would work only if the original query were really a *verbal* question – a question like *Is the English word "science" properly applicable to a hypothesis that makes reference to God?* But that was not the question: the question is instead *Could a hypothesis that makes reference to God be part of science? That* question cannot be answered just by citing a definition. (Plantinga 1997, 146)

At one level it is easy to answer Plantinga. It would indeed be very odd were I and others simply trying to characterize "science" as something which, by definition, is based on a (methodologically) naturalistic philosophy and hence excludes God, and then simply leaving things like that. Our victory (if denial or exclusion of Christianity is our aim) would be altogether too easily won. We would indeed simply be ruling religion out of science by fiat. But this is not quite what is happening. There is no attempt to offer an analytic definition of what one means by "science," as one might offer an analytic definition of "straight line" as meaning "shortest distance between two points." This is a definition which is analytic or stipulative. What is going on – what I was trying to do in Arkansas – is the offering of a lexical definition; that is to say, we are giving a characterization of the use of the term "science." What Plantinga in the passage quoted above calls giving an answer to a "verbal" question. And the suggestion is simply that what we mean by the word "science" in general usage is something which does not make reference to God and so forth, but which is marked by methodological naturalism. Moreover, whether one likes this fact or not, it is true. Since the scientific revolution, the professional practice of science has been marked by an ever-greater reluctance to admit social or cultural beliefs, including those of religion.

Plantinga may promote Augustinian science as "science," but it would be he who was making a stipulative definition.

There is another level to what is going on, and here one has to confess to a certain sympathy for Plantinga's complaint. He is surely correct that if all we had was a dispute about words, then the debate would be trivial. Whether one is defining "science" stipulatively or lexically, in itself nothing much rests on it. But there is more. The fact is that, having set the boundaries to science, many do go on immediately to claim that what lies beyond the boundaries is wrong or misguided or nonsensical. In the language we have been using, whatever people may say that they are doing – and many are proudly open in their actions – there is often a slide from methodological to metaphysical naturalism. The logical positivists used to claim that everything outside logic and science is meaningless, and this would certainly include Christianity. Plantinga is absolutely right that there is a tendency to characterize science on the basis of subjects like Darwinism and then to denigrate everything which does not fit the pattern. But note that this is surely only a tendency, and if one is indeed a committed Christian then there is nothing in Darwinism, or in the notion of science that it supports, which says that your commitment is wrong or stupid. Yours is not a scientific commitment, but you knew that already. If scientists and philosophers persist in saying that your position is meaningless simply because it is not science, then it is they who are guilty of arbitrary stipulative definitions.

Plantinga would spurn this proffered help. He switches from simply pointing out that, whatever the logic of the situation, Darwinians do slide into atheism. Now he takes a more philosophical tack, showing that methodological naturalism is more than just a heuristic. It forces you into some unacceptable ontological conclusions. If you are a Darwinian you are committed to the rule of law, and this enters into your definition of science. But, complains Plantinga, this means that your science cannot accept or treat of the unique or nonrepeatable. Laws are universal. And this means that at some level you are denigrating or dismissing the unique and nonrepeatable as beyond the reach of respectable or legitimate explanation. Which means that again Christianity suffers, because virtually by definition it deals precisely with the unique and nonrepeatable. Jesus was not just another prophet. He was the Christ, and uniquely he suffered, was crucified, and rose from the dead. So you can say what you like. At some deep level, Darwinism is anti-Christian.

Again there is a ready response. On the one hand, one can draw attention to the distinction between cosmic history and salvation history. Even if the story of Jesus be a story of unique events, this is salvation history – a religious story – set against the background of law-governed repeatable events, cosmic history – a scientific story. On the other hand, repeatability and uniqueness are somewhat relative terms. In a sense, everything is unique and, in a sense, everything is repeatable. Take the demise of the dinosaurs at the end of the Cretaceous period. This was in itself a unique phenomenon and unrepeatable; but, uniqueness notwithstanding, the demise was made up of many factors which can individually be brought under lawful understanding. It seems most probable that the event was triggered by an asteroid or a comet or some such thing hitting the Earth (Alvarez et al. 1980). This was no unique phenomenon, nor was the hitting of the Earth by the asteroid or comet such that the normal laws of nature – that is to say, Galileo's laws of motion – could not be applied. It is believed that there was a huge dust cloud raised, and the Earth became dark. Again, even if this was a unique phenomenon – and the dust cloud in the last century after the explosion of Krakatau makes one doubt this – one can still apply laws. One has all sorts of experience of dust causing darkness; then of darkness cutting off photosynthesis of plants and of the consequent dying of plants; and then of the consequent starvation of animals, which are part of the ecological food chain depending on plants. In other words, although the dinosaurs existed only once and will never reappear – so their demise was certainly something unique – the various components involved in the extinction of the dinosaurs are such that they can be brought under regularity. In principle, we have nothing different from any frequently repeatable phenomenon, like the death of annual plants at the end of every growing season.

Conversely and interestingly, one might also say that the Christian story is made up of many nonunique elements. Great healers do have a dynamic energy to bring people back from the brink, restoring them to health. Men and women of immense charisma do have the ability to fill groups with love and compassion towards each other, in ways barred to the rest of us – just as the Hitler-featuring Nazi rallies showed that such people can also turn crowds to vile and wicked ends. And it is not unknown for downcast groups to be filled with hope and joy and renewal, even in the face of the worst circumstances. Whether or not one thinks that law-breaking miracles were involved in this drama, the real point of this Christian story is that these phenomena all came together in a unique and overwhelming pattern. It is the story of order and meaning. Hence, in a way Plantinga helps us to see that science (including Darwinism) and religion (including Christianity), even if different, can come together in shared patterns of understanding. Because of its components and the way they are ordered, the Cretaceous extinction is not just any old event. And, because of its components and the way they are ordered, Christianity is not just any old event.

Plantinga is not finished. Harking back to some of the themes we discussed in an earlier chapter, he claims that if you are a Christian, then you ought to accept that the Jesus story involves miracles outside the bounds of law. But this acceptance is weakened if you deny a similar status to the miracles – including the creation story miracles – of the Old Testament. Hence, you should accept these earlier miracles also, and Darwinism obviously goes against this. This is why we need Augustinian science rather than Darwinian (naturalistic) science. We need an approach which recognizes God's actions as basic and not just incidental.

Natural laws are not in any way independent of God, and are perhaps best thought of as regularities in the ways in which he treats the stuff he has made, or perhaps as counterfactuals of divine freedom. (Hence there is nothing in the least untoward in the thought that on some occasions God might do something in a way different from his usual way – e.g., raise someone from the dead or change water into wine.) (Plantinga 1997, 149)

Plantinga's position is that, properly understood theologically, God's interventions and the running of law are a seamless whole of the same logical type. Therefore, from a Christian theistic point of view, there is absolutely no reason to deny the possibility of miraculous interventions. Indeed, Plantinga's position is that, as a Christian, one ought to expect God to be intervening – not out of a failure to do the job properly in the first place, but because God is always sustaining His creation. "God is already and always intimately acting in nature, which depends from moment to moment for its existence upon immediate divine activity; there is not and could not be any such thing as his 'intervening' in nature" (149).

What can one say here except that Plantinga is mistaken in thinking that his arguments here drive him inexorably towards a denial of Darwinism? Apart from ignoring the whole order-of-grace line of thought,

Plantinga's urge to miracles shows he has already made a commitment to what many would find as an unacceptably literal reading of the Bible.

> The issue, be it noted, is not whether God *could* have intervened in the natural order; it is presumably within the power of the Being who holds the universe at every moment in existence to shape that existence freely. The issue is, rather, whether it is antecedently *likely* that God would do so, and more specifically whether such intervention would have taken the form of special creation of ancestral living kinds. Attaching a degree of *likelihood* to this requires a reason; despite the avowed intention not to call on Genesis, there might appear to be some sort of residual linkage here. In the absence of the Genesis narrative, would it appear likely that the God of the salvation story would also act in a special way to bring the ancestral living kinds into existence? It hardly seems to be the case. (McMullin 1993, 312)

One final argument by Plantinga. He claims that the Darwinian scientist has pragmatic as well as theoretical reasons to be against Christianity. Many scientists, including Darwinians, would defend the naturalistic philosophy which lies behind their science, because even though there are gaps in our scientific understanding – the origin-of-life question for the Darwinian, for instance – enough solutions have been found in the past that one would be foolish to deny that they will continue to be found in the future. This will be pressed on the Christian, who will be encouraged (to say the least) to take a liberal attitude on miracles, interpreting them as explicable by the laws of nature and so forth.

Plantinga challenges this. While he agrees that giving up on methodological naturalism is in some sense what he calls a "science stopper" – something which brings methodologically naturalistic science to an end – as Christians, we have no reason to think that such science-stopping events do not happen. "The claim that God has directly created life, for example, may be a science stopper; it does not follow that God *did not* directly create life. Obviously we have no guarantee that God has done everything by way of employing secondary causes, or in such a way as to encourage further scientific inquiry, or for our convenience as scientists, or for the benefit of the National Science Foundation" (Plantinga 1997, 152–53). Perhaps, indeed, naturalism is forcing people away from interpretations and understandings which truly ought to be foundational elements of their religious beliefs. "Clearly we cannot sensibly insist in advance that whatever we are confronted with is to be explained in terms of something *else* God did; he must have done *some* things directly. It would be worth knowing, if possible, which things he *did* do directly; to know this would be an important part of a serious and profound knowledge of the universe" (153).

Again one feels a certain sympathy for Plantinga at this point. There is no question that many scientists, Darwinians at the front, take their naturalism so seriously (dare one say, religiously) that they sound like David Hume at his most ferocious. They simply would not accept any law-breaking miracle, and if indeed Christianity depends on them, so much the worse for it. Against this, I would note what has already been noted: that this is not a new argument uniquely inspired by Darwinism. One may be reinforcing it, but Hume did make the point, and he was a century before Darwin. More importantly, I would remind Plantinga that our powers of sense and reason are God-given, and although employing them in finding out the nature of the world may lead to revisions in our faith, this activity can hardly be considered inherently irreligious. Plantinga, a Calvinist, should be sensitive to this, for it was above all Calvin himself who stressed the Christian duty to discover the world the Creator has wrought. Science stoppers are just that: science stoppers. One might well say that they should be no more acceptable to the Christian than to the Darwinian.

Naturalism Self-Refuting

We are not yet finished with Plantinga. In an argument which has achieved some considerable notoriety, he has been claiming that Darwinism collapses in on itself, either in contradiction or in a circle which is intellectually stifling. Hence no Darwinian can be a true Christian, for as such an evolutionist one is forced into a radical scepticism which cannot be reconciled with genuine religious commitment. Everyone, and most especially the Christian, has to take a stand outside Darwinism (Plantinga 1991a, 1993).

The argument is as follows. If naturalism (in some sense) be true, then we should be evolutionists, Darwinian evolutionists of a kind. If we are evolutionists, then this must extend to our reasoning and cognitive powers. But, Darwinian evolution cares nothing for truth, only for survival and reproductive success. Hence, there is really no reason why our reasoning and cognitive powers should tell us the truth about the world. They just tell us what we need to believe to survive and reproduce, which information (although effective) could as easily be quite false. Plantinga tells the story of an overly rich dinner in an Oxford College, where Richard Dawkins spoke up for atheism before the philosopher A. J. Ayer – a classic case of preaching to the converted, I should have thought – and then goes on to draw a philosophical moral. Perhaps none of our thoughts can tell us about reality. Perhaps we are like beings in a dream world:

Their beliefs might be like a sort of decoration that isn't involved in the causal chain leading to action. Their waking beliefs might be no more causally efficacious, with respect to their behavior, than our dream beliefs are with respect to ours. This could go by way of pleiotropy: genes that code for traits important to survival also code for consciousness and belief; but the latter don't figure into the etiology of action. It *could* be that one of these creatures believes that he is at that elegant, bibulous Oxford dinner, when in fact he is slogging his way through some primeval swamp, desperately fighting off hungry crocodiles. (Plantinga 1993, 223–4)

Everything we believe about evolution could be false, and if that is not a reductio of naturalism, and anathema to the Christian, nothing is! Plantinga (rather cutely) refers to this as "Darwin's Doubt" because the aged Darwin himself expressed worries of this kind. "With me the horrid doubt always arises whether the convictions of man's mind, which have been developed from the mind of the lower animals, are of any value or are at all trustworthy. Would anyone trust in the convictions of a monkey's mind, if there are any convictions in such a mind?" (Plantinga 1993, 219, quoting Darwin 1887, 1, 315–16). Plantinga does not mention that Darwin immediately excused himself as a reliable authority on such philosophical questions, but no matter.

Candidly, I am not sure how seriously we are supposed to take Plantinga's argument and example. It is certainly true that even the most ardent Darwinian agrees that there are times when organisms and their characteristics will be out of adaptive focus. Genetic drift would be a case in point, as would that phenomenon mentioned by Plantinga, "pleiotropy," where a single gene controls two different characteristics and a nonadaptive feature piggybacks on an adaptive feature. But in no sense does Plantinga describe a situation remotely like the way that evolution truly works. Something like drift has minor effects – effects so minor (by definition) that they can slip under selection. Thinking that you are boozing it up with Freddie Ayer is not the way to fight off crocodiles. Fighting crocodiles requires force and defence and cunning and fear and split-second reaction to danger and much more. Ask any student of African mammals. At the very most, things like drift and pleiotropy and so forth are going to make one a bit uncertain about some things occasionally – when those things do not much matter anyway. If we need to know the truth, and we do need to know the truth when faced with crocodiles, drift will not stand in our way. It is not sufficiently powerful to make evolution through selection that maladaptive or ineffective.

There is a more general point which is worth making here. Behind his silly example, Plantinga has seized on a question of some importance. Quite apart from nonadaptive causes, deception for selection-related reasons does occur. Indeed, organisms are sometimes greatly deceived about the world of appearances, and this includes humans. Sometimes we are systematically deceived, as instructors in elementary psychology classes delight in demonstrating. However, such cases are really not so mysterious or inexplicable. Darwinian evolutionary theory gives good reasons why we are deceived. Why, for example, do we believe in the objective necessity of causal connection, even though, as Hume showed, there really is nothing there? Simply because those proto-humans who associated fire with burning survived and reproduced, and those who thought it was all a matter of philosophy did not. It was in selection's interest to make us think that causes really do exist as entities, and so we do.

But, and here is the key issue, we can work out that systematic misconceptions occur and that they come from selection because we do have reliable touchstones against which to measure them: that falling trees hurt, for instance; that drinking arsenic kills; that other people's genitalia (factoring in sex and orientation) are a sexual turn-on; that the grass really is green on a bright sunny day. These are not cases of misconception. There is no good reason why we should be misled by the hurtfulness of falling trees. The world is not crazy. We do not come up against wildly counterintuitive phenomena like Plantinga's example of fighting crocodiles while you think you are in Oxford deep in worthy discussion. The deceptions of selection work for good reasons, alongside the nondeceptions. In this sense, we could not be deceived all of the time. And we can use the general truthfulness of evolution to ferret out the instances of deceitfulness.

I doubt that this commonsense response will be the end of matters. Plantinga will again press his argument. Are we not being a little naive in thinking that our touchstones are reliable? How do we know we are not being deceived all of the time? Perhaps what we take to be the truest of the true, even genitalia, are having us on. Plantinga gives the example of being in a factory where everything seems to be red. The outsider knows that it is all a question of filters: unbeknownst to themselves, the workers are wearing spectacles which colour their perceptions. In truth nothing is red. But the man in the factory has no genuine touchstone by which to make a judgement, so no matter what care he takes in his assessments he will be mistaken. Perhaps we are in the same situation. Perhaps everything in the world/factory is false/red-seeming-but-not-really? To which I can only reply that it may be the case that we will never know the whole story and may be mistaken about any detail – although I do hope that genitalia never forsake us, at least in my lifetime. But I still claim that in real life we cannot be mistaken about all of the details. The factory example breaks down as an appropriate analogy precisely because at some point someone gets out to find that the redness is filter-caused. Unfortunately, we can never check that everything is true-seeming-but-not-really. We can never get outside our evolution-produced bodies. So in a way it is difficult to know precisely what one might mean by saying that our thoughts are all totally mistaken.

None of this is to deny or downplay this very important fact that all our knowledge starts with sensation and understanding and reasoning done with evolution-produced adaptations. And one consequence of this is that one cannot get directly in touch with what philosophers often call "metaphysical reality": the kind of ultimate being (or Ultimate Being) which exists whether anyone is around to sense it or not. (The tree which falls in the forest when no one is around, in the popular example.) But apart from the fact that there are serious doubts about whether this kind of reality truly makes sense, ignorance of its nature hardly tips one into the kind of radical scepticism that Plantinga threatens. The Darwinian simply denies that truth can mean correspondence between one's ideas and reality, arguing rather that truth means (as I have been arguing in the preceding paragraphs) a coherence between all the parts that we hold important and significant. Unless challenged, one accepts the touchstones and tries to make a comprehensive, consistent, and meaningful overall picture (Ruse 1986a).

As you might expect, Plantinga anticipates such a move and is scornful. For him, coherence is circular – you are justifying one part in terms of another, and then doing the whole process in reverse. Plantinga argues that a circular argument is still circular even if your premises take you all around the universe, before you end up back here with the conclusion. Here, I will simply deny that the circularity of coherence is vicious. Rather, as the success of science (including evolution) shows, you get an ever-bigger and better picture, as you (that is, the human race) get ever-more experiences and put them into the picture. In a way, "feedback" would be a better term than "circular."

In any case, the kind of coherence about which I am talking is hardly one to make Christian faith impossible. As Descartes showed in his *Meditations*, logically one cannot exclude the possibility of an evil demon who is corroding our most certain of beliefs, even those of mathematics and logic! If the Christian says that he or she has a direct line to God – faith or some such thing – the rest of us still have the right to ask whether this direct line is any more secure than any other. And the answer is surely that it is not. I am not now saying that one cannot be a Christian, whether or not one is a Darwinian. I am saying that we all start with ourselves and our powers and abilities. Plantinga is being naive or arrogant if he truly thinks that the Christian has an impregnable foundation of belief not shared by those of us who start from empirical evolutionary premises.

The basic problem is that we are not really arguing. Ultimately, nothing is going to make Alvin Plantinga sympathetic to Darwinian theory or to its underlying naturalistic philosophy. If he knew of the success of the science and the power of the philosophy, he would realize how his Christian faith cannot remain static and untouched but must move forward creatively to meet the challenges. As it is, he does his religion no great service.

CHAPTER SIX

Design

There is no standard Christian position on the role of reason in religion. Catholics think that "natural theology" has a significant and full role to play: "Illumined by faith, reason is set free from the fragility and limitations deriving from the disobedience of sin and finds the strength required to rise to the knowledge of the Triune God" (John Paul II 1998, 43). While there are Protestants who accept and even welcome natural theology, the "neo-orthodox" (like Barth) think not only that it fails but also that it is pernicious in its effects and promises. A true faith needs no proofs and indeed is destroyed by such proofs. Our radical freedom to accept God's gift of grace would be compromised were it possible to give logical proof of Christian claims.

Obviously we must discuss the interaction of Darwinism and natural theology, but equally obviously the Christian's own stand will have to be considered in any overall assessment of these issues.

The Teleological Argument

Arguments for the existence of God lie at the heart of natural theology. Some such arguments touch but slightly or not at all on the Darwinian system. The "teleological argument" or the "argument from design," however, is right on the front line. Many people, Richard Dawkins most vocally recently, claim that here above all Darwinism and Christianity come into conflict, precluding belief in both systems. By going back in history, let us see why this opinion might be held.

Notwithstanding Hume's criticisms – pointing to conclusions that he himself was not prepared to accept in full – the argument from design flourished right through to the nineteenth century. Interestingly, its most important base was Protestant Britain rather than Catholic Europe, mainly because – given the nonprofessional status of British science as opposed to that found on the continent, in France especially – British scientists had to work particularly hard to justify their activities to the outside (nonscientific) world (Appel 1987). Burnishing the argument from design was a perfect antidote to the worry that studying nature might put undue pressure on tenets of revealed religion. Its most famous formulation occurs in *Natural Theology,* by Archdeacon William Paley in 1802:

I know of no better method of introducing so large a subject, than that of comparing a single thing with a single thing: an eye, for example, with a telescope. As far as the examination of the instrument goes, there is precisely the same proof that the eye was made for vision, as there is that the telescope was made for assisting it. They are made upon the same principles; both being adjusted to the laws by which the transmission and refraction of rays of light are regulated. (Paley [1802] 1819, 1)

A watch demands a watchmaker. Hence an eye demands an eye maker – or rather, an eye designer. Call this "God": the God of the Christian, moreover, since the eye and other organic characteristics attest to a designer of great skill and power.

The popularity of this argument makes understandable one of the most important points about Darwinism: the author of the *Origin* accepted completely and utterly the initial premise of the teleological argument, namely that organisms are designlike (Ruse 1979a). Indeed, this is the problem to which natural selection speaks: the explanation of adaptations like the eye and the hand. It is here that Darwinism distinguishes itself from almost all other evolutionary theories. Darwin argued that, thanks to natural selection, we will have the formation or evolution of features like the hand and the eye, those very organs of which the natural theologians made so much. Darwin regarded the features as adaptations, as did the theologians. They were not just idle bodily parts or appendages, but things with a purpose or end or function. This is the reason that the *Origin* incorporates all of the teleological language of the theologians. If you like, put it this way: the metaphor of design is just as much a feature of Darwin's *Origin* as it is of Paley's *Natural Theology*.

Does Darwin Exclude Real Design?

Now what does all of this imply? Some people think that Darwin spelt the end to the argument from design. Before Darwin, one had no choice but to accept a Designer. After Darwin, the Designer was finished and the way was open for atheism.

> Paley's argument is made with passionate sincerity and is informed by the best biological scholarship of his day, but it is wrong, gloriously and utterly wrong. The analogy between the telescope and the eye, between watch and living organism, is false. All appearances to the contrary, the only watchmaker in nature is the blind forces of physics, albeit deployed in a very special way. A true watchmaker has foresight: he designs his cogs and springs, and plans their interconnections, with a future purpose in his mind's eye. Natural selection, the blind, unconscious, automatic process which Darwin discovered, and which we now know is the explanation for the existence and apparently purposeful form of all life, has no purpose in mind. It has no mind and no mind's eye. It does not plan for the future. It has no vision, no foresight, no sight at all. If it can be said to play the role of watchmaker in nature, it is the *blind* watchmaker. (Dawkins 1986, 5)

Because he did not know about evolution through selection, Hume hesitated before the final leap into nonbelief. Now such a leap is nigh obligated: "Although atheism might have been *logically* tenable before Darwin, Darwin made it possible to be an intellectually fulfilled atheist" (Dawkins 1986, 6).

But surely the Christian has a counter to this? One might argue that although selection makes redundant – closes off the option of – an intervening and designing God, it still leaves open the option of God's designing at a distance. Perhaps God put His design into action through the medium of unbroken law. Indeed, as Baden Powell argued in the years just before the *Origin*, perhaps a God who works this way is superior to a God who has to intervene personally and miraculously: "Precisely in proportion as a fabric manufactured by machinery affords a higher proof of intellect than one produced by hand; so a world evolved by a long train of orderly disposed physical causes is a higher proof of Supreme intelligence than one in whose structure we can trace no indications of such progressive action" (Powell 1855, 272).

Dawkins will have none of this. He regards Darwinism not simply as proving that the argument from design does not work, but as proving that atheism is true. Natural selection explains adaptive complexity. God simply cannot do this, because apart from anything else, one would then have the burden of explaining God.

Any God capable of intelligently designing something as complex as the DNA/protein replicating machine must have been at least as complex and organized as that machine itself. Far more so if we suppose him *additionally* capable of such advanced functions as listening to prayers and forgiving sins. To explain the origin of the DNA/protein machine by invoking a supernatural Designer is to explain precisely nothing, for it leaves unexplained the urging of the Designer. (Dawkins 1986, 141)

Dawkins is slipping in a strange premise, that complexity needs greater complexity to explain it. The whole point about reductionism is that one explains the complex in terms of the simple. But no matter. We can give Dawkins some of what he wants, but we are not obligated to give all. It is true that Darwinism shows that the need for an intervening designer is redundant. More than this. If you accept Darwinism, you reject the intervener. However, if you insist that the design demands a designer, then it is still open to you to accept that God did the job. More likely, if you accept God already, it is still very much open to you to think of God as great inasmuch as He has created this wonderful world. "What believers who have furnished such proofs [for the existence of God] have wanted to do is to give their "belief" an intellectual analysis and foundation, although they themselves would never have come to believe as a result of such proofs" (Wittgenstein 1980, 86). Even the neo-orthodox might go this far. There is nothing yet which stops the Darwinian from being a Christian.

But are we not being a little unfair to Dawkins at this point, missing the real force of his argument? His basic objection is that whether you think that God designed through miraculous intervention or through the medium of natural selection, you are still leaving unexplained the very existence and nature of this wonderful God who is supposedly capable of doing all of this. Which point of course is true and in the opinion of many is a good reason for nonbelief. Ultimately, assuming the existence of God really solves and explains nothing. Yet this surely is a problem for Christian belief generally and not something brought on by Darwinism specifically. There are of course various responses one can make to the problem, which may or may not be judged adequate. For instance, traditionally, God is thought to exist necessarily, so the question of His beginnings is ruled irrelevant. To which critics object that the idea of necessary existence is a conceptual confusion. At which point we can pull back gracefully and let the disputants argue among themselves. Their premises have nothing to do with evolutionary theory. Dawkins has not shown that being a Darwinian denies, or even exacerbates the difficulties of, Christian commitment. In the spirit of Baden Powell, one might think that God's magnificence is confirmed as one realizes that He does so much with so simple a mechanism as natural selection.

Is Selection Adequate?

Switch things around for a moment. We have been assuming that selection can do the job. But what if it cannot? What if there are aspects of the living world that in some sense, even in principle, Darwinism simply cannot explain? Does anyone truly think that Darwinians will show the appropriate modesty, retiring from the field and letting others move in? Surely not! Such aspects will be played down or denied or treated as unreal problems in the first place. And does this not mean that we then stand in danger of ignoring or denying or belittling aspects of the living world that, for the Christian, ought to be very important indeed? Because of our Darwinism — confident that it can, that it must, explain all — might we not turn away from precisely those things which theologically are the most significant?

This is the fear which underlies the thinking of biochemist Michael J. Behe, author of *Darwin's Black Box: The Biochemical Challenge to Evolution* (1996), a man who believes that he has made a breakthrough where "[t]he result is so unambiguous and so significant that it must be ranked as one of the greatest achievements in the history of science. The discovery rivals those of Newton and Einstein, Lavoisier and Schrödinger, Pasteur and Darwin" (232–3). Perhaps so, but moving to the arguments, let us see why he gives us reason to fear Darwinism. Behe's key notion is something he labels "irreducible complexity." Some organic phenomena are just so complex that they cannot have been produced by blind unguided law. That is just a fact of nature.

By *irreducibly complex* I mean a single system composed of several well-matched, interacting parts that contribute to the basic function, wherein the removal of any one of the parts causes the system to effectively cease functioning. An irreducibly complex system cannot be produced directly (that is, by continuously improving the initial function, which continues to work by the same mechanism) by slight, successive modifications of a precursor system, because any precursor to an irreducibly complex system that is missing a part is by definition nonfunctional. (39)

Behe adds, surely truly, that an irreducibly complex biological system has to be a major challenge to a Darwinian mode of explanation. Darwinism insists on gradualism, and this is precisely what is not on offer. "Since natural selection can only choose systems that are already working, then if a biological system cannot be produced gradually it would have to arise as an integrated unit, in one fell swoop, for natural selection to having anything to act on" (39). Which essentially means that natural selection is redundant.

As a matter of fact, Behe does not want to rule out a natural origin for all irreducible complexities, but we learn that as the complexity rises, the likelihood of getting things by any indirect natural route "drops precipitously" (40). As a physical example of an irreducibly complex system, Behe instances a mousetrap: something with five parts (base, spring, hammer, and so forth), any one of which is individually necessary for the mousetrap's functioning. It could not have come into being naturally in one step, and it could not have come about gradually. Any part-piece would not function properly alone, and any part missing would mean failure of the whole. It had to be designed and made by a conscious being – a fact which is true also of organisms. "The purposeful arrangement of parts" (193) is the name of the game.

Irreducible Complexity Challenged

Now what are we to say about this claim? Obviously, if Behe's overall argument is well taken, then Darwinism is in trouble and will surely strike back at Christianity. But are we to accept Behe? As it happens, Behe's choice of a mousetrap as an exemplar of intelligent design is somewhat unfortunate. All sorts of parts can be eliminated or twisted and adapted to other ends. There is no need to use a base, for example. You can just attach the units directly to the floor, a move which at once reduces the trap's components from five to four. But even if the mousetrap were a terrific example, it would hardly make Behe's point. No evolutionist ever claimed that all of the parts of a functioning organic feature had to be in place at once, nor did any evolutionist ever claim that a part used now for one end must always have had that function. Ends get changed, and something that was introduced for one purpose might well take on another purpose. It might be only later that the new purpose gets incorporated in such a way that it becomes essential.

Against the mousetrap, take the example of an arched bridge, with stones meeting in the middle and with no supporting cement. If you tried to build it from scratch, the two sides would keep collapsing as you started to move the higher stones into the middle. What you must do first is build an understructure, placing the stones on it. Then, when the stones are pressing against one another in the middle, you can remove the understructure. It is now no longer needed; although, if you were not aware that it had once been there you might think that it is a miracle that the bridge ever was built. Intermediate positions were impossible. Likewise in evolution: some pathway (say) exists; a set of parts sits idle on the pathway; then these parts link up; and finally the old pathway is declared redundant and removed by selection. Only the new pathway exists, although without the old one the new one would have been impossible.

Let us move now from analogies and pretend examples (though my own example is not so pretend if origin of-life researchers are right about the second stage of the Oparin-Haldane hypothesis [Cairns-Smith 1985]). We find that Behe's case for the impossibility of a small-step natural origin of biological complexity has been trampled upon contemptuously by the scientists working in the field. It is not just that they disagree, but that they think his grasp of the pertinent science is weak and his knowledge of the literature curiously (although conveniently) outdated. Take that staple of the body's biochemistry, the process by which energy from food is converted into a form which can be used by the cells. Rightly does a standard textbook refer to this vital organic system, the so-called Krebs cycle, as something which "undergoes a very complicated series of reactions" (Holum 1987, 408). This process, which occurs in the cell parts known as mitochondria, involves the production of adenosine triphosphate (ATP), a complex molecule which is energy-rich and which is degraded by the body as needed (say, in muscle action) into another less rich molecule, adenosine diphosphate (ADP). The Krebs cycle re-makes ATP from other energy sources – an adult human male needs nearly 200 kg a day – and by any measure, the cycle is enormously involved and intricate. For a start, nearly a dozen enzymes (substances which facilitate chemical processes) are required, as one subprocess leads to another.

Yet the cycle did not come out of nowhere. It was cobbled together out of other cellular processes which do other things. It was a "bricolage." Each one of the bits and pieces of the cycle exists for other purposes and has been co-opted for the new end. The scientists who have made this connection could not have made a stronger case against Behe's notion of irreducible complexity had they had him in mind from the first. In fact, they set up the problem virtually in Behe's terms: "The Krebs cycle has been frequently quoted as a key problem in the evolution of living cells, hard to explain by Darwin's natural selection: How could natural selection explain the building of a complicated structure in toto, when the intermediate stages have no obvious fitness functionality?" (Meléndez-Hervia et al. 1996, 302). What these workers do not offer is a Behe-type answer. First, they brush away a false lead. Could it be that we have something like the evolution of the mammalian eye? Primitive existent eyes in other organisms suggest that selection can and does work on proto-models (as it were), refining features which have the same function. Probably not, for there is no evidence of anything like this. But then we are put on a more promising track:

In the Krebs cycle problem the intermediary stages were also useful, but for different purposes, and, therefore, its complete design was a very clear case of opportunism. The building of the eye was really a creative process in order to make a new thing specifically, but the Krebs cycle was built through the process that Jacob (1977) called "evolution by molecular tinkering," stating that evolution does not produce novelties from scratch: It works on what already exists. The most novel result of our analysis is seeing how, with minimal new material, evolution created the most important pathway of metabolism, achieving the best chemically possible design. In this case, a chemical engineer who was looking for the best design of the process could not have found a better design than the cycle which works in living cells. (302)

Behe's knowledge of evolution is suspect. His knowledge of his own area of science is suspect. And the same is true when he moves into philosophy and theology. The common complaint about evolutionary theory is that it cannot be properly checked. The critics claim that it is too flabby to yield testable predictions. It is in some sense unfalsifiable (Popper 1974). But whether or not this is true (I do not happen to think it is), such a complaint must certainly be made of Behe's theory. How can you ever tell when irreducible complexity can be explained by evolution and when it must be explained by something else (or Something Else)? Behe himself admits that there is no sharp line, and he gives no real answers to this problem. Newton and Einstein and those other great scientists to whom he likens himself produced work which did lead to quantification and to measurement and prediction. As it stands, Behe's ideas can easily be protected against any counterevidence. You can explain some phenomenon through evolution? Then either the phenomenon was not irreducibly complex, or it was not complex enough. You cannot explain some phenomenon through evolution? Then either the phenomenon is too complex for an evolutionary explanation, or you will later find such an explanation. Heads I win, tails you lose.

More than this, there is a major unsolved problem about the way or ways in which intelligent design is supposed to act. Is it something built into nature from the first? If so, where is the quarrel with the Darwinian, for presumably laws had to effect the design, and why should not the designer work through natural selection? If the design is not in nature from the first, then was it added all at once to a primitive cell or does it occur piecemeal as needed? Without absolutely committing himself, Behe floats the idea that the design occurred all at once, asking us to suppose "that nearly four billion years ago the designer made the first cell, already containing all of the irreducibly complex biochemical systems" discussed in his book "and many others" (227). But if everything was done all at once, long ago, then how can it be (since the irreducible complexity of the higher animals and plants was not then needed) that it did not degrade or get eliminated, by random mutation or drift or selection weeding out the unneeded? If piecemeal, then whether or not it was put into play through a straight miracle or through a special kind of guided law, why do we have the evidence of Darwinian evolution (as for the Krebs cycle)? Why does the designer throw around such misleading clues? We are back with the logic of Philip Gosse, author of *Omphalos*. This is not very plausible, as science or as religion. (For much more on Behe's science, see Miller 1999.)

Intelligent Design

Qua scientist, Behe is careful not to identify his designer with the Christian God, and deliberately I have been saying nothing about this Being that Behe invokes. After all, our question is not whether one can be a Christian but whether being a Darwinian stops you from being a Christian. But if Behe's argument actually points away from the Christian God, this should be acknowledged, for then Darwinism is surely a more attractive alternative for the Christian. And this may indeed be the case. Let us suppose that a Behe-type designer does exist and is at work producing irreducibly complex organisms. Who then is responsible when things go wrong? We have all of the problems we have seen before. What about mal-mutations causing such awful things as Tay-Sachs disease and sickle-cell anemia? Is this just the fault of no one, or do we blame evolution? Why does the designer not step in here? It (let us not assume its sex) is pretty clever and could surely fix just one bad move. The whole point is that it can produce the irreducibly complex. So why does it allow – perhaps even produce – the not-very-complex-but-absolutely dreadful? Behe says that raising this problem is raising the problem of evil – How can an all-powerful, all-good God allow pain? And this is so. But labeling the problem does not make it go away.

There are some standard arguments addressing the problem of evil; we shall be starting that discussion in the next chapter. Here, although Behe himself is in as much trouble in the realm of philosophical theology as he was in the realm of biological science, let us see how others try to haul him from the hole into which he has pitched himself. The mathematician-philosopher William Dembski (1998a,b) recognizes that one must find some way to separate such things as mal-mutations from such things as highly complex functioning entities, else the whole new anti-Darwinian revival of the design argument (what its proponents call "intelligent design") comes crashing down. To this end, Dembski proposes something he calls an "explanatory filter." The essence of this idea is that you always explain things at the most economical or plausible level of understanding, and you only go on down to another level if the first level fails. So, faced with some (biological) phenomenon, you explain if you can through regular unbroken law. If that works, then the cheering can begin. Your job is finished. If it does not work, then you go to the next level: chance. If that works or is plausible, again your work is over. But if it does not work, then you must go on to another level: design.

The nice point is that there is no need to attribute to God all of the messy, unpleasant aspects of organic life. The beak of the finch on the Galápagos islands is clearly something produced by natural selection, and so, with such a law-based explanation, your job is finished. A mal-mutation is a random phenomenon – it is not something predictable within the context of Mendelian genetics – and so it is chance. It is inexplicable by law, but not such as to require further understanding. The origin of life cannot be explained by law, and it was certainly not chance. Here a design hypothesis is appropriate. And see how everything is kept clean and separate. You cannot blame God for mal-mutations. These are pure chance. "To attribute an event to design is to say that it cannot plausibly be referred to either law or chance. In characterizing design as the set-theoretic complement of the disjunction law-or-chance, one therefore guarantees that these three modes of explanation will be mutu-ally exclusive and exhaustive" (Dembski 1998a, 98).

A nice solution, but wrong. At the most charitable, there is a radical confusion between the meanings of "law," "chance," and "design." They are simply not "mutually exclusive and exhaustive" categories in the way that Dembski supposes. Fisher, the greatest evolutionist of this century – and, as arguably the greatest statistician ever, surely one who knew about these things – ran all three together! He believed that mutations come individually by chance, but that collectively they are governed by laws (and undoubtedly are governed by the laws of physics and chemistry in their production) and thus can provide the grist for selection (law) which produces order out of disorder (chance). He cast the whole picture within the confines of his "fundamental theorem of natural selection," which essentially says that evolution progresses upwards, thus countering the degenerative processes of the Second Law of Thermodynamics. And then, for good measure, he argued that everything was planned by his Anglican God! Remember, we are still living in the sixth day, "probably rather early in the morning" (Fisher 1947, 1001).

Returning to our worry, as soon as one has invoked design, at whatever level, then surely one can and should go back and reexamine attributions of chance (and law, for that matter). "Chance" is not a thing or an objec-tive entity. It is a confession of ignorance. My winning the lottery was a chance event, but this is hardly to say that it was an event outside of law – the laws of physics as the counters tumbled in the drum – and if God can create life, then He is surely up to seeing that I can get a million dollars that I did not earn or merit. So, it could all be part of His design. In short,

Dembski's help is no true help, and Behe is no better off than before. If God is directly responsible for the origin of life, or for the Krebs cycle, then He cannot escape responsibility for mal-mutations.

The sad truth is that Behe is in the same boat as those physicists we dismissed earlier. He has offered us a freshened-up version of the old "God of the gaps" argument for the Deity's existence. A Supreme Being must be invoked to explain those phenomena for which I cannot offer a natural explanation. But such an argument proves only one's own igno-rance and inadequacy. It tells us nothing of beings beyond science. In the words of the Christian theologian and martyr Dietrich Bonhoeffer: "We are to find God in what we know, not in what we don't know" (1979, 311).

Mind and Matter

Let us move next to the relationship between mind and matter. Again we find an argument purporting to show the inadequacy of pure Darwinism, and again we have a point where the Darwinian may be tempted to counter in such a way that there is a denial or belittling of something that the Christian can and should take as significant for belief. John Pol-kinghorne (1994) takes note of the isomorphic relationship between the facts of nature and the beliefs of mind. Snow is white, and we believe that snow is white. This is not very exciting, but this is just the beginning. What really impresses Polkinghorne is the way in which the human mind is able to transcend the vulgar and empirical and to inquire into the deeper mysteries of nature: theoretical physics, higher mathematics, and more. Surely, he argues – and as a theoretical physicist Polkinghorne is certainly qualified to argue here – this ability is proof of a designing, caring Mind which lies behind human intellectual activity. Indeed, a purely natural explanation cannot explain the correspondence between mind and theory and, if it tries, can do so only by undercutting the evidence of God's power and glory.

Confirming the Christian's worst fears – Polkinghorne is also an Angli-can priest – the Darwinian does certainly have a ready answer to this kind of argument. Let us first push the answer through to its end, and then ask about its implications for our overriding question: Can a Darwinian be a Christian? Simply, the Darwinian's claim is that the coincidence between mind and matter is indeed no chance, but that there is little need to suppose outside interference. It is just that physics and mathematics are adaptations forged by natural selection to enable us to survive and reproduce (Ruse 1986a; Bradie 1986). Leaving aside those special cases (discussed in the critique of Plantinga) where selection systematically deceives, the proto-human who realized that falling rocks tend to fall rather than rise up into the air survived and reproduced in a way that his less calculating cousin did not. The proto-human who did not realize that she was getting only two shares for the three she gave out, did less well in life's survival and reproductive stakes than she who was less gullible. There is no magic to science and mathematics. It is all in the genes. If you want to believe that everything adds up to Christian design, then you are free to do so; but there is no compulsion on the Darwinian, in this respect. The Christian should not make too much of what is going on here. "Creatures inveterately wrong in their inductions have a pathetic but praise-worthy tendency to die before reproducing their kind" (Quine 1969, 126).

Polkinghorne has an obvious response. While this counterargument might seem plausible for elementary physics and mathematics, can it possibly be adequate for more advanced areas of the subjects? Can it possibly be the case that evolution has anything to do with our grasping of the fact that space is non-Euclidean or that $e^{\pi i} = -1$? And these today are fairly simple concepts. The biggest mystery here is that Darwinians are so blinded by their theory that they cannot see how limited and limiting it truly is. And just putting things this way, even the hard-line Darwinian has to agree that there does seem to be a major gap. But there are a number of points which bear on the case.

First, no Darwinian is claiming that grasping $e^{\pi i} = -1$ has a direct bearing on survival and reproduction. The point is that mathematical and scientific claims are put together from simple claims in simple steps, and these basic units of knowledge and methodology are rooted in biology. Consider Euclid, for instance. One might plausibly argue that the axiom that shortest distance between two points is a straight line is Darwinian-based, even if one might doubt that the Pythagorean theorem is Darwinian-based. Second, note how mathematics and physics today are both necessarily limited in certain respects. Gödel's theorem shows that there are unprovable claims even in elementary mathematics. Would a Good God of the hands-on variety have left these dangling? If everything is contingent, then such undecidability is almost to be expected. Likewise, Heisenberg's Uncertainty Principle suggests that, even in theory, there are areas into which we may not go, questions we cannot answer. Again, this is all very surprising given a Good God; but it is to be expected if all science is a contingent outcome of the powers of beings forged by Darwinian evolution. I am not saying that this disproves God – one might think that this strengthens one's belief in a God who designs through evolution – but I am saying that it makes the all-powerful intervening designer less likely.

Third, it is surely open to the Darwinian to argue for some form of Platonism, at least with respect to mathematics. And here we are no worse off than, and perhaps even parallel to, the Christian. Where does the Christian (dissatisfied with the evolutionary proposal) think that mathematics resides? In the Mind of God, presumably. But what precisely does this mean? One supposes that there is a transcendent world, an ultimate reality in which the mathematical propositions in some form hold eternally. This is Plato's world of Forms or Ideas. "The Christian vision places the Forms securely in the Word of God" (Ward 1998, 107). I really do not see why a Darwinian should not hold to the Platonic vision as much as a Christian. The Darwinian already agrees that there is a world of physical reality, which may or may not have an ultimate explanation. Why should the Darwinian not also hold that there is a world of nonphysical reality, which likewise may or may not have an ultimate explanation? And if this world exists, why should not Darwinism open the doors? As Plato himself pointed out, once we are in, then a lot of non-Darwinian hard work will be needed to go from room to room; but that is another matter. As with empirical science, natural selection gives the necessary tools.

Polkinghorne raises a serious question. Darwinians today can hardly pretend that they have a full understanding of how adaptations forged through natural selection have become so powerful as to be able to grasp higher mathematics or quantum mechanics. There is much work to be done – but not by giving up on Darwinism because it is seen as a threat to religious belief. Although, in truth – and now we can return to our main question – need we see in any of this a threat to Christian belief? There is no longer a proof of God's existence, but is the mystery and wonder of higher mathematics any less now than it was before? Puny primate though I may be, I find the beauty and elegance of $e^{\pi i} = -1$ as moving as a Bach cantata, and I suspect for much the same reasons: reasons which reside in abilities given to me by evolution through natural selection.

Darwinism Explaining Christianity

I conclude this chapter by considering an argument which goes the other way. Could it not be that the Darwinian approach to function and design really does prove too powerful for the Christian? Could it not be that Darwinism shows that religion itself is just a part of the adaptive design of human nature, and that once we recognize this it will be seen that religion, including Christianity, falls to the ground?

This is certainly the position of Edward O. Wilson (1978). Wilson does not want to belittle religion in the fashion of Dawkins. He sees it as an important and significant aspect of human culture. But he wants to turn precisely this importance and significance back on itself. For him, religion exists purely by the grace of natural selection. Those organisms which have religion survive and reproduce better than those which do not. Religion gives ethical commandments, which are important for group living. Also, religion confers a kind of group cohesion, something which is a very important element of Wilson's picture of humankind. "A kind of cultural Darwinism . . . operates during the competitions among sects in the evolution of more advanced religions. Those that gain adherents grow; those that cannot, disappear. Consequently religions are like other human institutions in that they evolve in directions that enhance the welfare of the practitioners" (Wilson 1978, 174–5). Although Wilson writes here about cultural evolution, in fact he thinks that religion is ingrained directly into our biology. Thanks to our genes, it is part of our innate nature. "The highest forms of religious practice, when examined more closely, can be seen to confer biological advantage. Above all they congeal identity" (Wilson 1978, 188).

Religious enthusiasm is part of the human condition. We can explain religion. We can never eliminate it. At best, we can promote biology as an alternative secular religion: "The final decisive edge enjoyed by scientific naturalism will come from its capacity to explain traditional religion, its chief competition, as a wholly material phenomenon. Theology is not likely to survive as an independent intellectual discipline" (Wilson 1978, 192).

Explaining Religion Away?

Wilson's writings are rooted as much in his own childhood experiences of fundamentalist Baptism in the American South, as in any knowledge or study of empirical reality (Ruse 1999). But, taking his position at face value, let us ask about its implications for Christianity. In Wilson's own mind, what is happening is that Darwinism is explaining religion (including Christianity) as a kind of illusion: an illusion which is necessary for efficient survival and reproduction. Once this explanation has been put in place and the illusion exposed, one can see that Christianity has no reflection in reality. In other words, epistemologically one ought to be an atheist. Since Wilson still sees an emotive and social power in religion, he would replace spiritual religion with some kind of secular religion. That secular religion, as it turns out, happens to be Darwinian evolutionism. A Darwinian cannot be a Christian, but a Darwinian should be a Darwinian! We are dealing with a "myth"; but, when all is said and done, "the evolutionary epic is probably the best myth we will ever have" (Wilson 1978, 201).

Wilson's line of argument is hardly new. Karl Marx and Sigmund Freud proposed similar arguments – trying to offer a naturalistic explanation of religion, arguing that once one has this explanation in place, one can see that the belief system is false (Hick 1970). But is the inference in general well taken? And even if it is well taken, what of the specific case of Darwinism and Christianity? At the general level, it is certainly true that sometimes an explanation of why someone holds a belief suggests that, with respect to truth, the belief is not particularly well taken. Consider, for instance, the instance of spiritualism, particularly as it pertained to people's beliefs and practices during the First World War. Many bereaved people turned to spiritualism for comfort. And indeed, they derived such comfort, for they heard or otherwise received messages from the departed. However, all of us would now agree that, even in those cases where no outright fraud was involved, it was unlikely that the dead soldier was in fact speaking to those remaining. Peoples' strong psychological desires to hear something comforting led them to project and receive the desired messages, and so they heard them. Once one offers this explanation, seeing how unreasonable it is to expect that the departed were in fact speaking, the whole spiritualist position collapses.

Yet, not all explanations of why or how we get to believe things are necessarily such as to debunk the veracity of the belief systems. Suppose, for instance, one gives a scientific explanation of sight, showing how it is that someone is able to spot a speeding train bearing down on them. The fact that one can give an explanation – in terms of the eye's physiology and

of light rays and so forth – in no sense demotes or discredits the belief
that a speeding train is indeed bearing down (Nozick 1981). If anything, it
strengthens the belief. The question we must ask is whether religion is
more like the spiritualism case or more like the speeding train case – and
it is surely pertinent to note that this is a question which is neither asked
nor answered by Wilson. This omission does not mean that Wilson's
preferred explanation for religion – spiritualism rather than train – is
wrong. But it is to say that some additional argument is needed.

This incompleteness is a general feature of arguments like that of
Wilson – as it is, indeed, of those of Marx and Freud before him. They are
arguments that, to a certain extent, come after the event rather than
before. One becomes convinced that religion, let us say Christianity, is in
some sense inadequate or false. Then, one is led to ask exactly why it is
that people are led to believe it, and one offers some kind of materialistic
or naturalistic argument in response. This response in itself is not sufficient to show that the belief is false; at least, one needs some further
information as to why the response itself shows the belief to be false. And
this applies to the particular Wilsonian case of Darwinism and Christianity. The missing elements in Wilson's case are crucial. The fact that
one has an evolutionary explanation of religion is not in itself enough to
dismiss the belief system as illusory or false. More is needed.

It is true that people have proposed arguments suggesting that belief
in Christianity is unsound, ridiculous even. There are all sorts of para-
doxes which the Christian must face. But whether or not one can defend
Christianity against such charges, the charges themselves have not been
brought on by Darwinism, which is the nub of this discussion. Take the
problem of the Trinity. How can God be three persons in one, at the same
time? How can God even be God the Father and God the Son? What was
God the Father doing when God the Son was on the Cross, crying out for
help? Perhaps one can deal with questions like these. Perhaps not. We
have had two thousand years of debate on the issue, and it was a major
reason for the split between Western and Eastern Christianity (Pelikan
1971–89). But this is not our problem: Darwinism is irrelevant. In short,
Wilson's Darwinism in itself does not prove the inadequacy of Christian
belief; rather, his Darwinism shows why one might have a Christian
belief, if evolution be true.

Try again. Could one not argue that Darwinism shows that there is
something wrong with religion, since Darwinism is indifferent as to the

form of religious belief? It is true that different beings might – and indeed do – evolve different ways of sensing the train's approach. One uses sight, another uses hearing. But the long and the short of it is that one is going to have to sense the train in some fairly reliable sort of way, otherwise one is going to be wiped out. Religion, however, might be effective in achieving group cohesion, even though it takes on very different forms: monotheism, polytheism, animism, and so forth. All of which suggests that, given this range of biologically adequate options, Darwinism is more corrosive of religious belief than one suspected at first.

The problem for the Wilsonian is that one can mount this argument without really bothering too much about evolutionary biology. We know full well that different people do have different religious beliefs. Some are Christians, some are Jews, others are Muslims, and so on. In other words, what we know already is that culture has led to different, sincerely maintained religious convictions. And I hardly need say that there are already those today who think the argument is significant and quite corrosive with regard to Christian belief, or indeed any specific religious belief. I hardly need say also that there are standard replies that can be offered. One can suggest that one belief is better than others. Or one can argue that perhaps there is some common core to all religious belief, and that this is what counts. And note that as with the main argument, these counterarguments have little to do with Darwinism. For all that there are important issues here, Darwinism is not relevant to the case. Christian belief is being judged by other factors.

The conclusion is clear. Christians surely ought to consider seriously the empirical claims that Wilson and fellow thinkers are making about their religion. The theological implications being extracted are another matter. No sound argument has been mounted showing that Darwinism implies atheism. The atheism is being smuggled in, and then given an evolutionary gloss. This is no good reason for giving a negative answer to our title question.

CHAPTER SEVEN

Pain

The biggest question of all for the Christian believer is the "theodicy" problem. If, as the Christian believes, God is omnipotent (all-powerful) and all-loving, then why evil? If He is all-powerful, He could prevent evil, and if He is all-loving, then He would prevent evil. Yet evil exists. How do we explain it? What light for or against does Darwinism throw on this issue?

Utility Functions

It is customary and convenient to draw a distinction between two kinds of evil: moral evil, that is, human-caused evil – Auschwitz   and physical evil, that is, the pain of natural processes – the child with sickle-cell anaemia. Open to the Christian are a number of standard responses. The most popular and powerful argument to explain moral evil – one which goes back to Saint Augustine – is that it is something resulting from human free will. God in His love gave humans freedom, and that meant the freedom to do ill as well as good. Auschwitz is the result of human choice, human sinful choice, and as such is something for which God is not responsible, nor is it something that (having given us freedom) God could now stop. This does not mean that God is indifferent to suffering. He gave His life on the Cross to save us from our sins, and He suffers now with every evil act.

The adequacy of this particular argument depends crucially on our understanding of the notion of free will. Since Darwinism has significant implications for this particular topic, we will shelve our full discussion of moral evil until these are raised. Physical evil or pain will therefore be our concern here. And it has certainly been the case that Darwinians think that their theory is pertinent, showing the inevitability of such evil according to the Darwinian scenario and hence the implausibility of the Christian commitment. Free will cannot explain away the agony of the child in distress from a genetic ailment, nor even is it of much help to the prey of the predator.

Richard Dawkins has been particularly eloquent on this subject. He warms us up with a quotation from Darwin: "I cannot persuade myself that a beneficent and omnipotent God would have designedly created the Ichneumonidae with the express intention of their feeding within the living bodies of Caterpillars" (Dawkins 1995, 95, quoting a letter from Darwin to Asa Gray, May 22, 1860). Then he moves in for the kill, using something which he calls a "utility function" – that for which an organism or adaptation is intended. To work out a particular utility function you need to "reverse-engineer" the feature, to see what function or purpose is being satisfied or is intended.

Cheetahs give every indication of being superbly designed for something, and it should be easy enough to reverse-engineer them and work out their utility function. They appear to be well designed to kill antelopes. The teeth, claws, eyes, nose, leg muscles, backbone and brain of a cheetah are all precisely what we should expect if God's purpose in designing cheetahs was to maximize deaths among antelopes. (Dawkins 1995, 105)

The trouble is that, conversely, antelopes seem to have no less an effective utility function, namely that of escaping cheetahs! They are fast, agile, watchful, and so forth. Put together, and we have a violent conflict which seems designed only for one of those sensationalist televison programmes on African wildlife. Nature red in tooth and claw indeed. "What is He playing at? Is he a sadist who enjoys spectator blood sports?"

Dawkins's conclusion is not that there is an evil God or gods, or an ineffectual one, but rather that there is nothing. Nature is blindly indifferent. Cheetahs kill or starve and gazelles are ripped to shreds while still living. And it all means nothing.

In a universe of blind physical forces and genetic replication, some people are going to get hurt, other people are going to get lucky, and you won't find any rhyme or reason in it, nor any justice. The universe we observe has precisely the properties we should expect if there is, at bottom, no design, no purpose, no evil and no good, nothing but blind, pitiless indifference. As that unhappy poet A. E. Houseman put it:

> For Nature, heartless, witless Nature
> Will neither know nor care.

DNA neither knows nor cares. DNA just is. And we dance to its music. (133)

Explaining Pain

At one level, the appropriate response is that none of this is really the concern of the Darwinian as such. Pain and misery exist in the animal world, whether Darwinism be true or not. The creationist must accept that the cheetah hunts the antelope and that the parasite torments the caterpillar. The problem of physical evil is not something invented by the Darwinian. But in a sense, this is a cop-out. Whatever else, Darwinism certainly concentrates the mind on the problem of physical evil. It shows that it is not just some contingent thing, readily explained away. Rather, the way in which organisms were created and the way in which they function is one which necessarily entails a great deal of pain and suffering. There is no getting away from this or pretending that it is something which could be minimized or considered just an unhappy by-product of the evolutionary process. Pain and suffering are right there at the heart of things and are intimately involved in the adaptive process. No one is about to say that the antelope dying in agony in the cheetah's jaws is showing much adaptation in its suffering, but fear and pain clearly have their role. The burnt child fears the fire, and for good reason.

Moreover, pain and ill are involved in more than just the struggle for existence. The source of new variation, random mutation, as often as not causes pain and suffering. For every mutation which brings benefit, there are hundreds which spell doom and disaster. Yet again we have something absolutely central to the Darwinian evolutionary process. Randomness is the key to new genetic mutations, on which the Darwinian struggle depends to create selection and consequent adaptation. Darwinism is the antithesis to the theory of evolution through guided beneficial mutations. And worse than this. Selection can actually keep deleterious mutations "balanced" within a population, as a kind of price for good or healthy or advantageous mutations. Sickle-cell anaemia (caused by individuals having two sickle-cell genes) stays in the human population because carriers

with but one sickle cell gene have a natural immunity to malaria, an immunity not possessed by those with no sickle-cell gene at all (Ruse 1982).

Theodosius Dobzhansky was aware of the costs: "A species perfectly
adapted to its environment may be destroyed by a change in the latter if
no hereditary variability is available in the hour of need. Evolutionary
plasticity can be purchased only at the ruthlessly dear price of continuously sacrificing some individuals to death from unfavourable mutations" (Dobzhansky 1937, 126–7).

Pain as Illusory

Darwinism focuses on physical pain, a major stumbling block to Christian
faith. What does one say in response? The theologians have been at work
on the problem of evil far longer than the Darwinians. We cannot hide
behind them, but let us start by seeing what they have to say.

First, one might deny that the problem is genuine in some sense, for
one might deny that physical evil truly exists. There are various ways in
which one might try to do this: arguing that evil is not so much a positive
thing in itself, but rather an absence of something, namely good; or that
evil is a power of nothingness in some sense, a void in the creation, as it
were; or (perhaps somewhat along the lines of Mary Baker Eddy and the
Christian Scientists) that evil simply does not exist, that it is an illusion.
Perhaps one might take a leaf out of the Cartesian book, arguing that
animals at least are not thinking beings, perhaps not even sentient, and so
all of Dawkins's worries about cheetahs and antelopes are beside the
point.

This has never been a particularly convincing defence, and Darwinism
makes it even less convincing. Illusion or not, physical pain is very unpleasant for humans. Anyone who has had a close relative die of cancer
knows that. The fact that some things are psychosomatic – and most
things are not – does not make the suffering any less intense. In certain
circumstances, you may be able to think yourself out of pain, but often
you cannot, and even when you can it does not mean that the ill goes
away. The cancer sufferer overcoming pain still has cancer. And the
Darwinian qua Darwinian will have little time for the argument that the
problem of physical evil does not extend down to the world of the brutes.
Even if you accept that because of their intelligence humans have a
capacity for anticipating pain – including death – in a way that the animals

do not, as a Darwinian you will still see human suffering on a continuum with the rest of the organic world. It is unthinkable that there is a sharp break in kind between the dog cowering before the heated poker and the cautious human before the same poker, any more than that there would be a sharp break between the dog seeing the poker and the human seeing the poker. Human characteristics and features, adaptive or not, have a direct relationship with the characteristics and features of the animals. That is what Darwinian evolution is all about.

Pain as a Route to Faith

The second argument, couched in traditional terms, begins with the Fall. We are cast out, in a state of alienation from God. Sin is in our nature, pain is our destiny. Our way back is to reforge the relationship with God, through faith. But faith cannot be genuine if it is guaranteed (undermined, some would say) by reason and by tangible rewards. If there were no pain, we would not be drawn in our need towards God – the sensitive author of Job realized this – and if there were no pain, then faith would lose its meaning. It would simply be common sense that God is all-powerful and all-loving. There would be no gap to be bridged. "The Bible . . . proclaims the paradoxical possibility of faith and hope in God in spite of all evil and suffering. Some of us would even argue that faith has no intensity or depth unless it is a leap into the unknown in the face of such absurdity. Faith is always faith 'in spite of' all the difficulties that defy reason and science" (Haught 1995, 50). The English philosopher of religion John Hick, having rejected most of the traditional defenses of God against the existence of physical evil, writes: "The only appeal left is to mystery. This is not, however, merely an appeal to the negative fact that we cannot discern any rationale of human suffering. It may be that the very mysteriousness of this life is an important aspect of its character as a sphere of soul-making" (Hick 1978, 333–4).

This is a neo-orthodox position. Faith, to be faith, must be a leap into the absurd. Not entirely comfortably, Hick adds a twist to his case, arguing also that physical evil may be necessary for our moral development. Without it, we would feel no inclination to better ourselves in any way. "The systematic elimination of unjust suffering, and the consequent apportioning of suffering to desert, would entail that there would be no doing of the right simply because it is right and without any expectation of reward." Without random pain and suffering, we would always and only do good on the assured expectation of reward. There would never be the possibility or inclination to do good simply because it is good – to exercise what Kant (1949) called the "good will," acting virtuously purely for virtue's sake. But ultimately, we come back to mystery. "Suffering remains unjust and inexplicable, haphazard and cruelly excessive. The mystery of dysteleological suffering is a real mystery, impenetrable to the rationaliz-ing human mind. It challenges Christian faith with its utterly baffling, alien, destructive meaninglessness" (Hick 1978, 335).

I do not now judge this response in itself, but see how it meshes with Darwinism. And my sense is that the fit is a good one. The Christian defence does not take the fact of evil as just a coincidental given, but rather as a real presence in life: there, horrible, meaningless, depressing, defeating, challenging. In a similar fashion, as Dawkins shows powerfully, Darwinism stresses the natural evil in the world. It does not explain it, but it opens the way to the Christian response. If you can accept the Chris-tianity, then you can certainly accept the Darwinism. Conversely, if you are a Darwinian looking for religious meaning, then Christianity is a religion which speaks to you. Right at its centre there is a suffering god, Jesus on the Cross. This is not some contingent part of the faith, but the very core of everything. God is not some impersonal Unmoved Mover, who has little concern with the creation and who feels none of the joys and travails of the earthly creatures. God is not sitting on His backside in heaven, listening to one new Haydn quartet after another. God feels pain, physical and psychological, pushed to the limit that any of us can feel. There is the agony of the crucifixion and the despair of rejection: "My God, my God, why hast thou forsaken me?" Darwinism, a science which so stresses physical suffering, looks to Christianity, a religion which so stresses physical suffering and the divine urge to master it.

God Can Only Do the Possible

Third and finally in dealing with the problem of physical pain, there is a necessitarian argument based on the nature of law. God is free to create as He will, and because He is all-powerful and all-loving, He will create the best that He can. But this does not mean that God can do the impossible. God cannot make two plus two equal five. God can only do that which can be done, and the whole story of the Incarnation shows that what can be done is not defined simply by the laws of mathematics and logic. God wanted to save humankind from its sinful nature, but that did not mean that God could do it in any way He chose. He had to sacrifice Himself on the Cross. Likewise, God having decided to create, did then create – perhaps His choice, perhaps not – in an evolutionary fashion. And this being so, He was now locked into a path which would necessarily lead to physical evil. It comes with the method employed.

To see the force of this argument, let us suppose that God might have used "better" laws of nature, that is, laws which do not lead to physical evil. For instance, let us suppose that God made the laws of nature so that there would be no chance of humans or others being poisoned by the ingestion of dangerous substances. Unfortunately, mere tampering – fine tuning – would hardly be adequate.

Would not either arsenic or my own physiological composition or both have to be altered such that they would, in effect, be different from the present objects which we now call arsenic or human digestive organs? To change the actual world sufficiently to eliminate natural evil, and therefore to instantiate a possible world with different natural laws, would necessarily entail change in existing objects themselves. (Reichenbach 1982, 110–111)

Indeed, the changes might well have to be so drastic that no longer could one properly say that one still had the substances with which one started. "They would have to be different in some essential respects, such that with different essential properties they would become different things altogether" (111).

And this is just substances. Imagine what kinds of wholesale changes would be needed to pain-proof various processes. Fire could no longer burn, for fear that children and others might get trapped in smoke-filled apartments. But if fire did not burn, how could I warm myself through the Canadian winter and how cook my food and so much more? One change by God would require another and another and another, until everything had been altered. And could this be possible? Where would it end, and where could it end in a satisfactory manner?

In particular, the introduction of different natural laws affecting human beings in order to prevent the frequent instances of natural evil would entail the alteration of human beings themselves. Human beings are sentient creatures of nature. As physiological beings they interact with Nature; they cause natural events and in turn are affected by natural events. Hence, insofar as humans are natural, sentient beings, constructed of the same substance as Nature and interacting with it, they will be affected in any natural system by lawful natural events. These events will sometimes be propitious, and sometimes not. And insofar as man is essentially a conscious being, he will be aware of those events which are not propitious and which for him constitute evils. Therefore, to prevent natural evils from affecting man, man himself would have to be significantly changed such that he would be no longer a sentient creature of nature. (Reichenbach 1982, 111–112)

And even now, who dare say we humans would be better situated. "Whether humans would have evolved but no infectious virus or bacilli, or whether there would have resulted humans with worse and more painful diseases, or whether there would have been no conscious, moral beings at all, cannot be discerned" (113). The world is a package deal, and we simply have no right or authority to say that God could have created in such a way as to prevent such physical evil as there is. The hard nature of physical existence and being is not therefore a rebuke against an all-powerful God.

Almost paradoxically, the Darwinian supports this argument; and by a nice turn of fate, the strongest support comes from the arch-atheist Richard Dawkins himself! For the Christian, the key aspect of organic form is its adaptedness – you see God's glory in life's functioning – and if you take a Darwinian position, then this adaptedness or functioning is a major source of the pain and suffering that you see in the world. Darwinism equals natural selection, and physical or natural evil is a result of the causes or a consequence of this selective process. Here, therefore, we would seem to have a place where God might have done otherwise. Would not things be better all around had He got adaptedness by a much nicer physical process than selection? Dawkins (1983), however, argues strenuously that selection and only selection can do the job. No one – and presumably this includes God – could have got adaptive complexity without going the route of natural selection. Why is this so? At least partly because adaptation and its complexity simply could not be produced by most putative evolutionary processes: by saltationism – evolution by jumps – for instance. As a matter of empirical fact, hopeful monsters – viable new life forms which represent complete breaks with earlier life – simply do not exist in nature.

But Dawkins's claim is rather stronger than this. It is not just a question of hopeful monsters not existing. In some real physical sense, they could not exist – at least, they could not exist and have been produced by natural processes. "Wherever in the universe adaptive complexity shall be found, it will have come into being gradually through a series of small alterations, never through large and sudden increments in adaptive complexity" (Dawkins 1983, 412). The point is that physical processes do not suddenly and spontaneously bring about adaptive complexity. The only sudden changes are those that destroy or degrade. They are never creative. Boeing 747s crash into the ground and in an instant they are no more. Boeing 747s do not lie in pieces around the junkyard or on the ocean bottom and then in an instant form a fully functioning flying machine. In the case of organisms, there is no known physical rival to the slow, creative, adaptive-complexity-forming process of natural selection. So it is selection or nothing. "However diverse evolutionary mechanisms may be, if there is no other generalization that can be made about life all around the Universe, I am betting that it will always be recognizable as Darwinian life. The Darwinian Law . . . may be as universal as the great laws of physics" (Dawkins 1983, 423). You cannot get adaptive complexity without natural selection.

The Christian positively welcomes Dawkins's understanding of Darwinism. Physical evil exists, and Darwinism explains why God had no choice but to allow it to occur. He wanted to produce designlike effects – without producing these He would not have organisms, including humankind – and natural selection is the only option open. Natural selection has costs – physical pain – but these are costs that must be paid. And this applies also if you think that a scientific solution must be found to account for the appearance of humans, and if you think that arms races offer the most convincing explanation. The pain and violence which results from these is simply an inevitable tariff for achieving the desired end. What more can one say?

Well, one thing one might say is that one should beware of Greeks bearing gifts. The philosopher Daniel Dennett (1995) refers to natural selection as the "universal acid," meaning that once it is up and running it corrodes everything. Having made appeal to the ubiquity of natural selection, should we not now allow the strength of Dawkins's earlier-encountered worry that invoking God as an explanation of design is no explanation, because God in turn requires explanation? If design can come only through selection, does this not mean that God Himself had to be the product of selection? In which case, do we need another God behind our God, and back in regress? To borrow a thought from Johnathan Swift:

> So, naturalists observe, a flea
> Hath smaller fleas that on them prey;
> And these have smaller fleas to bite 'em,
> And so proceed *ad infinitum.*

As before, the answer will surely be that God's existence and nature is not subject to or in need of the explanation that the contingent objects of this world demand. God exists necessarily and is immune to all acids, no matter how corrosive. Less metaphorically, the Christian will say that God is creator, not created. Even if we agree that God necessarily creates and fashions through a selective process, this tells us nothing about His own nature and being. It certainly does not tell us that He had to be made through natural selection. The Dawkins-Dennett worry is thus without foundation. Darwinism does not dissolve away Christian belief.

Faith and Rationality

Of course, in speaking thus of God's nature, a necessary Being who exists eternally, the Christian admits – insists – that there is an element of mystery. We are speaking of a Being about whose essence we can be only dimly aware. Which point brings us back to the main topic of this chapter, the problem of physical pain. We have yet been silent on the consequentialist concern that, given the costs, God's ends could never justify the means. It is customary to draw a distinction between God's freedom of action in creation and His freedom after creation. Once God had set things in motion, leading to animals and ultimately to humans, He could not then prevent pain and suffering. His freedom now lay in His response to such pain or suffering: love for and empathy with the sufferer, rather than indifference or stern judgement at the way in which such suffering is so often self-inflicted. However, it has always been an essential claim of Christian theology that God's creation at the beginning of time was an entirely free and loving act. And here's the rub! Would a good God freely have created the universe and its life in the first place, knowing the inevitability of the pain and suffering? As Dostoevsky asked in *The Brothers Karamasov,* could eternal happiness really be worth the needless suffering of one small child? There are some mutations which have quite horrendous effects, causing the most devastating of physical and psychological illnesses. Young children compulsively mutilating themselves, for instance, wracked with pain, their mental deficiencies depriving them of any chance of happiness and a normal life. Is eternal bliss for any of us – that child, the Pope, Mother Teresa, you or me – worth the suffering of that child? Dare one say yes?

As with questions about God's ultimate nature and being, the Christian response is couched in terms of faith and mystery. "Revelation has set within history a point of reference which cannot be ignored if the mystery of human life is to be known. Yet this knowledge refers back constantly to the mystery of God which the human mind cannot exhaust but can only receive and embrace in faith" (John Paul II 1998, 14). Terrible as the natural occurrence of pain and suffering may seem, it is part of God's purpose embraced in His love, that will be revealed to us in the fullness of time. Now we see but "through a glass darkly." Then we will see clearly. "The mystery of suffering is great, but so is the mystery of persisting hope" (Polkinghorne 1989, 64).

This is the third time in this chapter that the question of faith and mystery has been raised in the face of serious conceptual objections. Is it not now time to ask if this is ever the appropriate response for the Darwinian? As a scientist, one is committed to reason and logic and the evidence of the senses, and to following through to the conclusions, however unwelcome they may be. But now one is being asked to pull back, to jettison one's rationality, in order to make room for belief. One must embrace a reason-defying faith. Can this ever be fair? Can this ever be a demand to make of a reasonable person? Is not this desertion of the rational the very thing one has committed oneself not to do? Dennett (1995) makes such a complaint: "You must not expect me to go along with your defense of faith as a path to truth if at any point you appeal to the very dispensation you are supposedly trying to justify. Before you appeal to faith when reason has you backed into a corner, think about whether you really want to abandon reason when reason is on your side" (154).

Dennett's gripe is that we all want to use reason and understanding and the evidence of the senses when the going is good, or when important practical matters are at issue. If we were falsely accused in court, for example, we would be very upset if the judge simply threw out the evidence in our favour and intuited the "truth." And we would think you slightly crazy if you went to a surgeon who was guided solely by a little voice from within. Why then, when the evidence points the other way, should we be able to abandon common sense and reason, appealing rather to faith and to mystery?

I know it passes in polite company to let people have it both ways, and under most circumstances I wholeheartedly cooperate with this benign arrangement. But we're seriously trying to get at the truth here, and if you think that this common but unspoken understanding about faith is anything better than socially useful obfuscation to avoid mutual embarrassment and loss of face, you have either seen much more deeply into this issue than any philosopher ever has (for none has ever come up with a good defense of this) or you are kidding yourself. (154–5)

As one professional philosopher listening to another, I find this an emotionally powerful riposte. Reason is what we live by. But is it quite the conversation stopper that Dennett takes it to be? Picking up on an argument made in the last chapter, notwithstanding the significance of reason, this century's findings in science and mathematics must surely have infused people with a little modesty about their ability to peer into the nature of ultimate reality. The electron, for instance, really does seem to have contradictory properties, and the only way we can avoid them is by ruling out questions which seem likely to demand answers in terms of such contradictions. That is what Heisenberg's Uncertainty Principle is all about. But avoiding problems is not exactly equivalent to finding out what is going on. And the same is true of other ideas in modern science. Often we can give a mathematical characterization, but again this is not exactly the same as having a full insight into what is happening out there.

The point is not that these scientific ideas are crazy ideas, anything but. Nor am I saying that they force you to be religious or prove the Trinity or anything like that. Rather, the point is that our understanding goes so far and no further. And, picking up on the argument from the last chapter, the interesting corollary is that the Darwinian is in a stronger position to explain and accept our limitations than the Christian theorist (like Polkinghorne), who thinks that our reasoning powers all come straight from God. This theorist has to introduce ad hoc hypotheses to explain why God has given us a power of reason which goes so far and no further. The Darwinian knows at once that our limitations come from our having evolved in certain ways. These are ways appropriate to our station, that is, midrange primates who came down out of the trees and went into the garbage and offal business. We do not as a matter of everyday practice think in quantum terms because our world is not a quantum world. A wave is not a particle. We do not think (to take another example) in non-Euclidean terms because our world is not a non-Euclidean world. Our perceptual space is not curved. We do think in food-fussy ways, because ours is a world of rotting meat. Stinking beef gives you a stomachache.

We can indeed stretch out from the familiar to the unfamiliar. That is what modern science is all about, and it is very successful. But we are stretching out – anyone in Darwin's time who thought in quantum terms would have been accused of disordered thinking – and that does mean that questions which make sense at our normal, evolved level of activity cannot be answered – perhaps should not be asked. "What do quarks look like?" is not only a question which cannot be answered but also a question which does not make a great deal of sense. And more than this. There is no guarantee that such stretching is going to be capable of infinite extension. We may just run out of steam, not because there is nothing beyond but because we cannot go further with what we have. Men can run a four-minute mile, an undreamed-of achievement for the Greeks. They will never run a two minute mile. That is beyond all dreams. Of course, we humans are inclined to think that we have an absolute lien on all of rationality, that anything beyond can either be ferreted out by us or is not worthy of attention; but that is because natural selection has done its job. If we spent our time worrying about reason and its limits, we would be no good at surviving and reproducing by our wits, which is the direction in which we have evolved.

This is our world, and as I pointed out in disputing with Plantinga, this is the best we do have and can have. You cannot escape it in any rational or reasonable way, even if you believe in God. But precisely because you (as a Darwinian) are working within the world as you can know it, you ought to show a little modesty about your limitations. I am not now saying that there is a world of contradictions beyond our ken. I stress again that I do not think you can flip straight from quantum mechanics to esoteric Christian theology. Gould (1999) is scathing about those who would try to understand the many-faceted nature of God in terms of particle/wave duality. I would not be so harsh. Metaphors are what we live by. What I do say is that as a Darwinian you ought to be dubious about thinking that your selection-based attributes and powers, including mental attributes and powers, give you total insight into ultimate metaphysical reality.

The point was well put in a famous passage by Haldane. Deliberately, he is taking a skeptical position; but it is surely a position friendly to the Christian.

Our only hope of understanding the universe is to look at it from as many different points of view as is possible. This is one of the reasons why the data of the mystical consciousness can usefully supplement those of the mind in its normal state. Now, my own suspicion is that the universe is not only queerer than we suppose, but queerer than we *can* suppose. I have read and heard many attempts at a systematic account of it, from materialism and theosophy to the Christian system or that of Kant, and I have always felt that they were much too simple. I suspect that there are more things in heaven and earth than are dreamed of or can be dreamed of, in any philosophy. That is the reason why I have no philosophy myself, and must be my excuse for dreaming. (Haldane 1927, 208–9)

It is a dream which we all must share. Dennett's complaint is important. People do blow hot and cold on reason as it suits them. With him, I have a deep hatred of irrationality: fake cures for cancer, lunatic religious sects. But, when one is dealing with a system like Christianity, which is grappling with some of the really deep issues which face humankind – physical pain and moral evil, particularly – precisely because one is a Darwinian, one ought to be sympathetic when the system runs into doubt and mystery. Being a Darwinian does not compel one to be a Christian; but, because one is a Darwinian one is opening the way for someone to be a Christian.

CHAPTER EIGHT

Extraterrestrials

"Are you a Whewellite or a Brewsterite, or a t'othermanite, Mrs Bold?" said Charlotte, who knew a little about everything and had read about a third of each of the books to which she alluded.

"Oh!" said Eleanor; "I have not read any of the books, but I am sure that there is one man in the moon at least, if not more."

"You don't believe in the pulpy gelatinous matter?" said Bertie.

"I heard about that," said Eleanor; "and I really think it's almost wicked to talk in such a manner. How can we argue about God's power in the other stars from the laws which he has given for our rule in this one?"

"How indeed!" said Bertie. "Why shouldn't there be a race of salamanders in Jupiter, why shouldn't the fish there be as wide awake as the men and women here?"

"That would be saying very little for them," said Charlotte. "I am for Dr. Whewell myself; for I do not think that men and women are worth being repeated in such countless worlds. There may be souls in other stars, but I doubt their having any bodies attached to them. But come, Mrs Bold, let us put our bonnets on and walk round the close . . ."

This exchange, just before a moonlit walk, comes in Anthony Trollope's novel *Barchester Towers,* published in 1857, two years before the *Origin* appeared. It was the topic of the hour. Did the truth lie with William Whewell, Darwin's philosophical mentor and author of the (barely) anonymous *Of the Plurality of Worlds: An Essay,* who argued that we humans are unique in the universe? Or did it lie with Whewell's great Scottish rival, Sir David Brewster, student of the theory of light, biographer of Newton, and author of *More Worlds than One: The Hope of the Philoso-*

pher and the Creed of the Christian, who argued that intelligent beings populate the whole universe, even the sun?

History of the Problem

The debate is an old one (Dick 1982). The Greek atomists had argued that space and time are infinite, and that populated worlds of intelligent beings exist no less infinitely. Aristotle, however, had put this earth uniquely at the centre of the universe. Beings could not have existed on the heavenly bodies, for these bodies were in the place of unchanging perfection. In any case, such beings would have fallen off and down to the earth at the universe's centre. This view was one which found favour with Christianity, for there was little support for extraterrestrials to be found in the Bible. However, in the sixteenth century with the coming of Copernicus, pluralism became a topic of plausible discussion, and in the succeeding centuries it became – if not orthodoxy – at least the most favoured viewpoint amongst the educated public. Nor were distinguished supporters lacking. Kant (1981), for instance, populated most of the solar system, not to mention the worlds beyond.

There were, however, theological tensions. One supporter of intelligent extraterrestrials – Giordano Bruno – was burned at the stake in 1600. Kepler put the two sides well, showing the stress that the subject caused between revealed theology and natural theology. On the one hand, intelligent beings not of this Earth seem to threaten our special relationship with God. "For if their globes are nobler, we are not the noblest of rational creatures. Then, how can all things be for man's sake? How can we be the masters of God's handiwork?" (Kepler 1965, 43). On the other hand, worlds without denizens seem a pointless exercise of the Almighty's power. "Our moon exists for us on the earth, not the other globes. Those four little moons exist for Jupiter, not for us. Each planet in turn, together with its occupants, is served by its own satellites. From this line of reasoning we deduce with the highest degree of probability that Jupiter is inhabited" (Kepler 1965, 42). The fact that natural theology generally won out over revealed theology – people simply could not accept that all of those distant universes were truly inert and lifeless – did not mean that folk were always entirely happy with their choice.

Organic evolution enters a new element into the equation, and Whewell saw this at once. If the doctrine be true – Whewell was not at this point reacting against Darwinism but against the evolutionism of the anonymously published *Vestiges of the Natural History of Creation* (which appeared in 1844 and was in fact authored by the Scottish publisher Robert Chambers) – then it would seem that you are going to have life just about everywhere, as a matter of course, and humans could be very far down the scale of progressive being. To quote the poet Alexander Pope, would we not find:

> Superior beings, when of late they saw
> A mortal man unfold all Nature's law,
> Admir'd such wisdom in an earthly shape,
> And shew'd a Newton as we shew an Ape.
> (*Essay on Man*, Epistle II, 31 34)

Is this really compatible with our relationship with the Christian God?

Apparently so! Most of the Victorians – with the notable exception of Alfred Russel Wallace (1903), natural selection's codiscoverer – simply could not see this as a real threat. They were all for evolutionary progress but could not imagine that it might lead to an organism significantly superior to a nineteenth-century Briton. They agreed with Brewster against Whewell that extraterrestrials do not threaten Christian belief. Today we have perhaps less confidence in the innate superiority of Western man. It is easier therefore to see that the canny philosopher truly knew a pertinent new factor when he saw one. Evolution as fact and theory was significant for the plurality of worlds debate and remains so today. Cutting right to the present, the questions for us are: first, where precisely does the Darwinian stand on this issue of extraterrestrials, and then second, what implications follow for Christianity? Let us take them in turn.

The Darwinian Case for Extraterrestrials

There are those who argue that (Darwinian) evolution implies not only the existence of extraterrestrials but also the likelihood that they will be much like us. Understanding a "humanoid" to be a "natural, living organism with intelligence comparable to man's in quality and quantity" (Simpson 1964, 259), Robert Bieri (1964) sees such beings as virtually the inevitable consequence of the evolutionary process. He makes much of the already-introduced notion that there are internal and external constraints upon the solutions that organisms can adopt to evolutionary challenges and that these lead to certain convergent paths:

The physical properties of the elements, the forms of energy available, and the environmental conditions which would allow life to arise and evolve are such that severe limitations are imposed on the number of routes available to evolving forms. The number of alternative possibilities is by no means infinite; on the contrary, the number is quite limited. This limited number of available routes has led to the innumerable cases of convergent evolution in plants and animals. (452)

Thus, in building your humanoid, Bieri argues that there is a very good reason why the mouth tends to come at the front of an organism and the anus at the back. It is anything but chance; there are good adaptive reasons. You get your food from outside, so you need to be on the lookout, with sensory organs geared up to find suitable fodder and with grasping organs at the ready to grab when so ordered. These sorts of things are obviously going to be at the front – their position virtually defines what you mean by "front." And, this being so, the last thing you need is a mouth somewhere around the back, out of the way and requiring special effort to stuff. The mouth needs to be in the thick of the action. And the same is going to apply to the brain, given the wiring which is needed to control the sensing and grasping apparatus. "The anterior mouth with surrounding sense and grasping organs has been independently evolved in group after group, again and again. It is not surprising therefore to find the largest ganglion of the brain at the front end in close proximity to the major sensing organs" (453). The anus, on the other hand, is expelling what you have already got and want to jettison. You can tuck it around the back. In any case, you need to keep it away from the food.

Again, we expect to find our humanlike being going about on legs rather than (say) slithering like a snake, and not an odd number of legs nor a large number either. Two seems like an ideal number. "It seems most probable that our extraterrestrial humanoid will have either two or three sets of paired appendages. I'm willing to bet on the smaller number" (456). Brains likewise. We are not like those huge dinosaurs with minibrains who needed a second control centre in their butts. Even the number of fingers probably falls into a restricted range. Perhaps one could go up to six fingers (in which case presumably we would count by twelves), but intelligent beings would be unlikely to have claws or some other type of grasping apparatus, and they would be no more likely to go up to digits by the dozen. All in all: "If ever we succeed in communicating with conceptualizing beings in outer space, they won't be spheres, pyramids, cubes, or pancakes. In all probability they will look an awful lot like us" (457).

The Case against Darwinian Extraterrestrials

Would Bieri find many evolutionary biologists following and accepting his arguments? My sense is that his kind of argument finds a lot more favour among cosmologists and others at the physical end of science, those interested in space travel or communication with extraterrestrials. Evolutionary biologists tend more towards the critical stance of the paleontologist G. G. Simpson, who authored a classic exercise in scepticism, "The Nonprevalence of Humanoids." His attitude is extremely negative: "There are four successive probabilities to be judged: the probability that suitable planets do exist; the probability that life has arisen on them; the probability that life has evolved in a predictable way; and the probability that such evolution would lead eventually to humanoids . . ." Simpson argues that the first probability is fair, the second much lower, the third vanishingly small, and the fourth effectively zero. "Each of these probabilities depends on that preceding it, so that they must be multiplied together to obtain the probability of the over-all probability of the final event, the emergence of humanoids. The product of these probabilities, each a fraction, is probably not significantly greater than zero" (Simpson 1964, 258–9).

Sounding like a forerunner of Gould, Simpson argues that it is chance and contingency all the way. You simply cannot bank on humans or humanlike organisms coming out at the end. "Both the course followed by evolution and its processes clearly show that evolution is not repeatable. No species or any larger group has ever evolved, or can ever evolve, twice. Dinosaurs are gone forever. Nothing very like them occurred before them or will occur after them." The same is absolutely and completely true of humans also. We have arrived here on Earth. There is absolutely no reason to think – and many reasons not to think – that humans will ever appear again on Earth or on any other planet. Mutations are random, you cannot expect them to repeat; environmental circumstances are unique and not to be found in the same order or magnitude elsewhere; and selection is always taking things off in different directions.

There is a more or less random element in evolution involved in mutation and recombination, which are stochastic, technically speaking. Repetition is virtually impossible for nonrandom actions of selection on what is there in populations. It becomes still less probable when one considers that duplication of what are, in a manner of speaking, accidents is also required. This essential nonrepeatability of evolution on earth obviously has a decisive bearing on the chances that it has been repeated or closely paralleled on any other plane. (Simpson 1964, 267)

Humans simply will not be repeated. Nor does Simpson think it very likely that humans themselves will go on evolving, so that we could expect there to be superintelligent humanoids throughout the universe. Using a version of the argument which haunted R. A. Fisher, Simpson points out that the poor and stupid tend to breed and have large families, whereas the rich and intelligent practice birth control and hence (in a way highly counterproductive from a Darwinian perspective) have much smaller numbers of offspring. Superintelligence is possible, but our experience is that it is not likely. "Future evolution could raise man to superb heights as yet hardly glimpsed, but it will not automatically do so. As far as can now be foreseen, evolutionary degeneration is at least as likely in our future as is further progress."

Assessing the Options

Flat disagreement. Is there any way in which we can achieve a resolution, if not a compromise? Probably not, in any absolute sense. We are dealing with one event here on Earth and trying to extrapolate into the unknown. Short of (highly unlikely) intergalactic space travel or (so far impossible) communication, there is bound always to be some element of doubt. However law-bound one might think evolution to be, one is truly dealing with something filled with chance and contingency. Making absolute inferences is highly risky, even if you are as unlikely to be found out as confirmed. For this reason alone, it would be silly to claim definitively that Darwinism has implications for extraterrestrials which make impossible Christian belief: or which make Christian belief mandatory, for that matter.

Our ignorance is not absolute. With respect to life itself, the Darwinian will probably agree that it does recur, perhaps often, in the universe. After all, the evolutionist does claim that the appearance of life on this Earth is no miracle. And if this be so, the possibility is open for it to have happened elsewhere on the literally billions of planets which must occur throughout space. Of course, this may or may not be particularly pertinent to Christianity. It is certainly not as pertinent as humanoids. But even with respect to humanoids, many Darwinians think we are not totally in the dark. Those who argue most strongly on this question are, to a certain extent, arguing past each other. So much depends on what you mean by "humanoid." If you mean "humanlike," then Darwinians are united in denying their frequent (or infrequent) possibility. But if you mean "intelligent being of some kind," then in fact the disagreements start to vanish. For those Darwinians who are progressionists – a group which includes Simpson, for that matter – even if humans never occur again, one might surely expect something with intelligence to reappear elsewhere in the universe. And indeed, for all that he is against human-oids, Simpson does seem to agree that something of this nature might exist elsewhere. As does Gould, incidentally. Remember: "I can present a good argument from 'evolutionary theory' against the repetition of any-thing like a human body elsewhere; I cannot extend it to the general proposition that intelligence in some form might pervade the universe" (Gould quoted by Dick 1996, 395).

All of this seems to be taking us back to a theme which we have encountered before. There are vacant ecological niches ("zones") waiting to be conquered – sea, land, air, culture – and intelligence is bound to emerge sooner or later. And even if the physical form is not the same, the intelligence is bound to follow familiar patterns.

The point I wish to stress is that again and again we have evidence of biological form stumbling on the same solution to a problem. Consider animals that swim in water. It turns out that there are only a few fundamental methods of propulsion. It hardly matters if we choose to illustrate the method of swimming by reference to water beetles, pelagic snails, squid, fish, newts, ichthyosaurs, snakes, lizards, turtles, dugongs, or whales; we shall find that the style in which the given animal moves through the water will fall into one of only a few basic categories. (Conway Morris 1998, 204–5)

The same applies to culture and thinking.

Not every evolutionist would be entirely happy with these confident conclusions. Richard C. Lewontin (1978) is one of those who question the whole notion of an empty niche waiting to be occupied. He thinks, rather, that organisms create their niches as much as they find them. He would doubt the existence of a cultural zone, out there waiting to be conquered by intelligence, on planet after planet. Conversely, others have wondered if there might not be another zone beyond culture, for something over and above thinking and intelligence. Simpson (1964) is pretty short with this suggestion: "There is not a scrap of evidence that 'life as we do not know it' actually exists or even that it could exist – evidence, for example, in the form of detailed specifications for a natural system that might exhibit attributes of life without the basis of life as we do know it" (255). Of course, even if there were life occupying new unexplored zones, it would not necessarily be the case that the denizens of such zones would be endowed with something "superior" to intelligence, whatever that might mean.

Negative Implications for Christianity?

Accept, if only for the sake of argument, that at least some Darwinians seem prepared, albeit hesitantly and with many qualifications, to allow the possibility of alien intelligence. Turn now to Christianity. What about the theological challenges posed by such extraterrestrial intelligence? At one level, this is all rather remote, conceptually as well as physically. People who feel their faith either challenged or supported by hypothetical aliens have probably been watching too much television. One feels sympathy for the rabbi who wrote, "Never before have so many been so enthusiastic about being so trivial" (Dick 1996, 502). But at another level, there are some real worries, and as was the case in the past, we find that opinion seems to be divided on this topic. There are those who think that extraterrestrials would challenge Christianity, and there are those who think that the fears are overblown.

Some of the concerns are traditional, but (as Kepler knew) none the less emotive for all that. One question which truly terrifies many is that if the universe is so vast, and if there are so many intelligent beings across the reaches of space, how possibly can God have concern for us humans, on this rather grubby little planet that we inhabit? It is bad enough with the population explosion here on Earth. That in itself waters down our relationship with God. Now we are talking about literally billions of souls which will need to be saved. It is all very well talking about God's knowing of every sparrow that falls. What about me, if He has got His mind focused on some little problem in the philosophical community on Andromeda?

Another concern centres on the question of intelligence. Let us agree that we need not worry too much about alternative forms of intelligence or something over and above intelligence, occupying a niche or zone beyond culture. Simpson's commonsense attitude assures us that there is really not much point in worrying about that anyway. But along with Alexander Pope, we may still fear that we are very much down the intelligence scale of galactic beings. What right have we to assume that we are the Newtons of the universe? Life as we know it has been around for nearly four billion years. Human intelligence dates from less than one million years ago. Human technology is but a few thousand years old, at the most generous. Surely there must be lots of beings who far exceed us in intelligence, and this being so, would they not be closer to God? "Being made in His image" has to involve intelligence in some form, and the more intelligent you are the more you are in His image. God may be worried about us, but perhaps more at the level of dogs and cats – or sparrows – than of really important beings. Come back in a few million years, when we have had the chance for a bit more progressive evolution, and then we can reassess the situation.

Third, even if we assume that there is a whole set of extraterrestrial intelligent civilizations, have we as Christians the right to think that God came to them in the form of the Christ, for their salvation? Is there not something unique about Calvary? Is this not a special one-time bond and act between a creator and His creatures? Or are we to suppose that, rather like a traveling circus, every Friday somewhere in the universe Jesus is being nailed to the Cross? Does this not reduce the Saviour to a stuntman, rather like a daredevil being shot from the barrel of a gun? In fact, things are even worse than this. The philosopher Roland Puccetti (1968), who has become a kind of modern-day Whewell, worries that if there are extraterrestrials (and he thinks that there are), and if they exist at the same time that we do (and he thinks this is possible), then they too might have need of a redeemer at the same time that we do. And, since Christ became flesh, we have the logically impossible demand that one person become two persons, at different places in the universe at the same time. We have our Christ here on Earth, and we have an "X-Christ" doing his thing many light years away across the universe. "We have Jesus Christ and they have 'X-Christ', both natures having been assumed by God and both species incorporated into Christ's 'manhood' and 'X-hood', respectively, as separate incarnations of the divine Word." This is an impossible situation. "If the son of God is numerically one how can He also be fully human and fully 'X' at the same time, i.e. two distinct corporeal persons? For two corporeal persons are not one. What is more, we should then have just as much reason to worship the 'X-Christ' as to worship Jesus Christ, and Christianity would embrace no longer a Trinity but a Quarternity" (139–140).

It seems the only escape from Puccetti's worries and those others given earlier would be to deny the existence of extraterrestrials, something the Darwinian will not necessarily want to do. Either that, or to admit that Christianity is simply a religion of this Earth – and only of a certain part of this Earth at a particular time, for that matter – which is certainly to downgrade its pretensions and presumed importance significantly. Not exactly a positive response to our title question.

Worries Overblown?

What can one say looking at matters from the side of religion? First, there is the worry that extraterrestrials water down our relationship with God. How can He care about us, if we are no longer unique but one of billions of creatures clamoring for His attention? This is the argument with the most powerful emotive force. Somehow, we humans seem to get lost – literally as well as metaphorically – in space. To which one can reply that, emotively powerful though the argument may be, it is not one which is new: Christians have been worrying about these sorts of issues since the Copernican revolution, when it was shown that the universe is far vaster than had hitherto been conceived. And this was but to repeat a fear of King David: "What is man that thou art mindful of him, and the son of man that thou dost care for him?" (Psalms 8:2)

Emotion aside, it is surely an integral part of Christian theology that God is indeed big enough to handle large and larger sizes and numbers and more. Remember that Christianity is rooted in Judaism, which back at the time of Abraham was practiced by just a very small band of nomadic people, and which certainly did not consider Yahweh or Jehovah to be the only god around. Judeo-Christianity has been expanding its scope ever since – first (thanks to Jesus and Paul) out to the Gentiles, and then (thanks to missionaries) out to all peoples of the world, a population which has been growing by leaps and bounds, especially in this century. There are far more humans today than there were at the time of Jesus. Are we to assume that God is not big enough to handle the population explosion? If so, perhaps someone should tell the Catholic Church, which is so adamantly opposed to artificial methods of birth control.

Let us make another point, to fight emotion with emotion. If evolution be true, then in some very real sense we humans are all part of one big family, no matter what our numbers. For the Christian, is this not the fulfillment of God's promise to Abraham? "I will make of you a great nation." A nation so great that Abraham is told, "Look toward heaven, and number the stars, if you are able to number them" (Gen. 12:2, and 15:5). Christians should not read this literally. Rather, as one can read the creation stories metaphorically – as telling us of God's relationship to humans and of our obligations to nature – so one can read this promise metaphorically, as referring to the family status of humankind. God tells the father of the tribe that there will be an ongoing vital group, descended from a common stem, and thus bound by blood ties. The numbers speak to the greatness of that group – "So shall your descendants be" – rather than to the insignificance of the individual. Is this not what evolution is all about? We are all, whatever our race or individual nature, part of the group, bound by common descent. Indeed, we are at one with the whole of living nature.

This does not speak directly to the issue of extraterrestrials; but even here there is comfort and emotional solace, if one sees Darwinism as but part of a larger evolutionary picture. The Big Bang suggests that, vast though the universe may be, it all came originally from one point. Hence, all of reality as we know it is related, going back to a common source. Development and evolution has been the story since, but ultimately everything stems from one root moment and event. If there is intelligent life throughout the universe, it (like us) is part of that grand tree of being. The magnitude of the universe and the possibility of billions of inhabitants chills the believer's frame. Evolution brings warmth.

Move next to the worry that with a host of extraterrestrials we are bound to be demoted in status, for some (perhaps many) of these beings will be far superior to us morally and intellectually. Surely, therefore, God will have them as His chief focus of attention. But against this, Christian theology is decisive. Are we really to suppose that God will not love us because we are not the brightest children in the universe? What sort of father is that? In any case, remember that we are not God. We are made in God's image. Even if there are brighter beings than we, humans have enough intelligence to exercise free will, obey God's laws, and worship Him as Creator of everything. That is what is needed for Christian practice, and what we have is quite enough. Let others, brighter beings, work out their own relationship with God.

Not that Darwinism is necessarily throwing obstacles in the way of Christian belief. Even if you do accept the controversial claims about the progressive nature of the Darwinian evolutionary process, this does not, as such, imply that organisms are going to go on evolving towards higher intelligence. There is more to being a successful organism than having a high IQ. Sociality counts too, for instance. Simpson notes that degeneration may be our fate. But there is more even than this. One might argue that a being with an intelligence such as ours is inherently unstable. Humanoids (or any intelligent beings) are likely to have a dark side – more on this in a later chapter, when we turn to original sin – and as soon as intelligence of our level is reached, nuclear (and biological and chemical) weapons and their attendant horrors are made possible. Dare anyone say, given what we know of ourselves in this century, that organisms of our intelligence would have the luxury of another two or three million years of nuclear- (biological-, chemical-) weapon-free existence, to engage in the struggle and selection – or eugenic breeding programmes, if you prefer – to produce yet more intelligent beings? I am far from convinced that, anywhere in the universe, beings of our intelligence can last for more than a few thousand years.

The Role of Jesus

Move now to the worries around the status and role of Jesus. What about the worry that new worlds will have missed the Incarnation – must miss the Incarnation – which was and had to be an event unique to our Earth? Can beyond-the-Earth intelligences ever be saved? If you are a liberal Christian you will find some way around this problem, because you will have tackled already the related problem of those humans born before Christ or born after him but never hearing of his existence. But what if you are more conservative, feeling that perhaps these unfortunates were damned (or at least not saved)? One move open here to the conservative Christian is that of challenging the presumption that intelligent extraterrestrials do need saving. The Incarnation was needed here on Earth because we are sinful, a result of past actions. Perhaps extraterrestrials all live in a state of purity, in the absence of original sin. The experience of this Earth is not necessarily a model for other worlds. We are not simply treating of a question of laws which must necessarily be extrapolated elsewhere. We are talking about actions performed by free beings. If there were no sinful Adam, there would be no need of a sacrificial Jesus.

As you might realize, an approach like this poses major problems for the more liberal Christian. The literal reading of Genesis raises all sorts of worries about interpretations of original sin as inherited guilt. If you add an evolutionary twist to your story, thinking that we and the extraterrestrials came about through natural processes, then you have additional difficulties with the assumption of an original Adam and so forth. I shall be discussing later ways of giving a Darwinian interpretation to the doctrine of original sin; but you may anticipate that they will not make more reasonable the solution just given. So let us move to another conservative solution, which escapes these worries. Suppose we accept that the extraterrestrials might be sinful. There is no need to hypothesize that Jesus has come or will come to save them!

This may conflict with our understanding of God's all-loving nature, but to the Calvinist such an attitude is unwarranted sentimentality. God's grace is His gift, not our right. The Darwinian above all others should recognize that some are saved and some are lost – selection is a law of nature as well as of God – and that while God's actions are not capricious (any more than are those of selection), success and failure do not come through intrinsic merit. Why should I have a good nature rather than my brother, any more than I should have some favourable variation rather than my sister? And why should my nature merit salvation, any more than my variation merits reproductive success? The plight of extraterrestrials is no more troubling than that of those already-mentioned unfortunates on this Earth who were never exposed to the Gospel message.

What about Puccetti and his worries that Christ could only come in one place and at one time? This was not something to worry the Church fathers. Christ did come, and there is an end to it. But if those Darwinians who support extraterrestrial intelligent life are correct, then should it worry us? Frankly, if indeed we did find Puccetti's X-Christ elsewhere in the universe, complete with X-Sermon on the Mount and X-Crucifixion, at a minimum we would have an all-new and powerful proof for the existence of God. Could such a repetition of the Earth experience – I would think, incidentally, that one might need an X-Saint Paul – be just a fluke? Even Richard Dawkins (or rather the X-Richard Dawkins?) would be scrambling to bolster up his thinking about cultural evolution and to show just why we might expect such a convergence of ideas and events.

Assuming now that the time would not be available for a succession of nonoverlapping appearances, what about the claim that the person of the Christ could not appear in more than one place at the same time? Ernan McMullin (1980) is particularly harsh with this argument: "There is an odd, ungenerous fundamentalism at work here, a refusal to allow for the expansion of concept, the development of doctrine, that is after all characteristic of both science and religion" (88). If Christianity can wrestle with such scientific advances as the Copernican and Darwinian revolutions – and McMullin believes that it can – then at the very least it is incumbent upon the believer in extraterrestrial intelligence to do likewise. McMullin takes umbrage at the literalism of Puccetti's objection that the concept of a person could not apply to God or to Jesus appearing simultaneously to beings in different parts of the universe. "These objections take on a certain irony when it is recalled that this concept [of a person] originated in the theological discussions of the fourth century centering around the Trinity and the incarnation. Theologians have always insisted on the analogical character of the concept of person and on the 'negativity' of our knowledge of God generally." McMullin concludes: "If a human 'person' cannot be in two places at once, does it follow that if God incarnate can be, the term 'person' is inapplicable to him?" (87)

This is a little hard on Puccetti. It is fair to ask somewhat more of the concept of "person" before we vanish into mystery and negativity. But McMullin is surely right to insist that the notion of a "person" as applied to God has to be analogical to the application to humans. Moreover, for the Christian the God-notion has to embrace the possibility of different forms and simultaneous appearances. The Trinity demands this. And then once you start talking about the Holy Ghost you are dealing with a Being that can manifest itself always throughout space and time. So extraterrestrials are hardly going to pose radically new problems in this respect. And this being so, given that the Christian presumably does find satisfactory the theological answers to God's personhood, we can bring this discussion to a close. Christianity, a fairly traditional form of Christianity, has resources to deal with anything thrown at it by extraterrestrials. Darwinism certainly does not obligate a belief in such beings, especially not of the intelligent variety. But if you follow those people – a group which certainly contains serious and eminent Darwinians – who think that biological science makes such beings plausible or even probable, there is no reason now to cast aside your Christian faith.

CHAPTER NINE

Christian Ethics

Alvin Plantinga's attack on Darwinian naturalism destructs through its own failings. But I do understand why he feels so strongly. He is defending his deeply held religious beliefs against what he clearly sees as another, rival, secular religion. And surely the comments we have seen from Edward O. Wilson alone should alert us to the fact that Plantinga's worries are not without foundation. Anyone who talks of replacing one "myth" with another has gone beyond the bounds of the purest science. Not that Wilson is the first evolutionist to try to make a metaphysics, a secular religion, from his science. He stands in a tradition which goes back to Charles Darwin and earlier. Indeed, one might say that this is evolution's oldest tradition. The very first evolutionists, men like Erasmus Darwin in England and Jean Baptiste de Lamarck in France, were open in their hope that evolution could substitute in some way for conventional religious beliefs.

In the middle of the nineteenth century, Thomas Henry Huxley, Ernst Haeckel, and (above all) Herbert Spencer set out to make of evolution a Christianity-substitute: a new world picture that could challenge and replace the old religions, one far more suited to the new industrial, urban, capitalist age than were the systems of the past (Ruse 1996a). Although Darwinism today is much more than an ideology, more than a new religion of humanism or naturalism or whatever, the tradition of so regarding it has persisted down through this century, many years after the publication of the *Origin*. Julian Huxley, grandson of Thomas Henry, was one of the most ardent enthusiasts in this respect, as is attested by his popular book *Religion without Revelation*.

Religions usually incorporate some kind of moral code or tradition – rules for proper conduct – and Christianity is a paradigm in this respect. Likewise for evolution-as-religion. It too is a font of moral prescription: most famously, so-called social Darwinism. Many people consider this an old, discredited nineteenth-century movement, but thanks particularly to Wilson, the past twenty years have seen more activity on the biological front than perhaps at any previous time. In this and succeeding chapters, my aim is to compare and contrast ethical thought and behaviour in the two domains: Christianity and Darwinism. First, I shall sketch the points of Christian ethics. Then I shall look both at traditional evolutionary ethicising and at recent work in this field, seeing the points of agreement and possible conflict between Christian thought and evolutionary (especially Darwinian) thought.

As I begin, let me remind you of – perhaps introduce you to – a distinction which is customarily and conveniently made in philosophical circles. This is the distinction between "normative ethics" (also known as substantive ethics) and "metaethics" (Taylor 1978). The former area of inquiry looks at the rules of proper conduct: "What should I do?" The latter area of inquiry looks at foundations for proper conduct: "Why should I do that which I should do?" Thus, normatively, Immanuel Kant (1959) asked that we follow what he termed the categorical imperative. In one formulation: treat people as ends rather than as means. Do not just use folk for your own purposes or benefit: regard them as worthwhile entities, in their own right. Never make an example of someone, simply for the sake of others. At the foundational level, Kant argued that (normative) morality is necessary and has its justification in the fact that no society of rational beings could function without it. It is not that there is some outside force or power to which you can and should appeal. It is rather, that a lack of normative morality would lead to civil chaos or what Kant termed a (social) "contradiction."

I am not commending Kant particularly, just using him as an example. We must keep in mind, in discussing both Christianity and Darwinism, that any adequate analysis of ethics must offer answers at both the normative level and the metaethical level.

The Gospels

Christianity rises out of the Jewish religion and was deeply influenced in its earliest years by Greek philosophy (Wogaman 1993). Both of these elements, the revelatory and the reasoned, can be found in Christian ethics. The Jews, of course, had their codes of proper behaviour. Some prescriptions were what we today would regard as customs, perhaps having tribal sanction or sanitary force, rather than the strictly ethical. One thinks, for instance, of the demand that males be circumcised and of the various dietary rules and restrictions. Some prescriptions were more directly ethical: the Ten Commandments are the prime example. However, the God of the Jews did not always order His People to act in ways that we today would find morally admirable. Indeed, towards alien peoples, He could show a ferocity which would make the average contemporary ethnic cleanser look positively harmless. But particularly as we come down towards the time of Christ, we sense a more enlightened and universalistic ethic. The love commandment ("love your neighbour as yourself") makes its appearance (Wallwork 1982; Betz 1985). Not that Judaism then or now was simply a pale proto-version of Christian thought. For the Jew, lifelong celibacy has always been something to be excused or explained away, rather than cherished for its own sake. For an embattled and threatened people, having a family is a positive obligation.

Jesus took Judaism, adapted it, and (Christians would say) transcended it. "Think not that I have come to abolish the Law and the prophets; I have come not to destroy them but to fulfill them" (Matthew 5:17). He was raised within the Jewish law, and it was the background to his thinking; but he could be casual or even callously indifferent towards its observance. He was little bound by Sabbath restrictions, for instance, and (although obviously one must take care in interpreting isolated comments) could be chillingly unsympathetic to the family ties and obligations of himself or his followers. Much more positively, Jesus wanted to go well beyond the limited reciprocation one finds characteristic of Jewish law – an eye for an eye and a tooth for a tooth – and (particularly as expressed in the Sermon on the Mount) wanted to extend moral behaviour out beyond the hitherto-marked outer bounds. One should not simply be restrained in the face of violence and unfair treatment: one should return hate with love. One should not simply give alms to the needy (the widows and orphans): one should give and give and give, until one has no more to give. One should not simply keep one's hands off the wives of others: one should not even lust after them in one's heart. One should not simply help those in one's own group: one should (as is shown by the parable of the good Samaritan) extend one's aid out to all people. One should not simply worship God: one should give up everything and follow Him. Question: "Good Teacher, what must I do to inherit eternal life?" (Mark 10:17) Answer: "Go, sell what you have, and give to the poor, and you will have treasure in heaven; and come, follow me" (Mark 10:21).

This is radical stuff indeed and, as scholars point out, must be framed in the context in which Jesus thought and lived and spoke (Ramsey 1950). His was an apocalyptic age, one that expected the Messiah and the coming judgement of God, and Jesus himself preached in this context. His human nature (as opposed to his divine nature) limited his understanding of God's plans, for clearly Jesus himself expected the end to come soon: within his own lifetime or at least that of his followers. Realization of his limited perspective may finally have come to him on the Cross, but his commands were directed towards hearers whom Jesus expected would soon be facing the end of time. For this reason, we do not find Jesus offering either a directly equivalent system to the Jewish law or a philosophical system as one finds in the writings of the great Greek philosophers. It is true that there are some dicta of practical importance, about divorce, for instance – not to mention his evading of the trap of sedition by advising his followers to render unto Caesar those things that are Caesar's. Generally speaking, however, in the preaching of Jesus we do not find an articulated moral system for ongoing societies: not even for those of yesterday, let alone for the technology-fuelled mega-groups within which we live today.

Developing Christian Ethics

It fell to Jesus' followers to develop and build an ethical system for societies which are going to persist and which are facing ongoing points of moral conflict, within and without. Saint Paul was the first to plunge right into this task, stressing the love commandment and offering counsel to the new and growing Christian communities within the Roman Empire. Yet although there is that in his writing which has inspired moral reformers for two millennia – "There is neither Jew nor Greek, there is neither slave nor free, there is neither male nor female; for you are all one in Christ Jesus" (Gal. 3:28) – at another level he was deeply conservative. Slavery as a social custom is accepted; the subordinate status of women is stressed; the immorality of homosexual activities is reaffirmed (unlike Leviticus, Paul includes lesbians explicitly in his prohibition); and we start to see the chilling attitude towards heterosexuality that marks out

Christianity from other great religions. (See especially I Corinthians 7.) Better to marry than to burn, but better not to have any sexual activity at all. Even touching women is proscribed. Of course, one should put all of this in context. Christian restraint was a welcome move in an era when sexual laxity was the norm. But a pattern was set.

The Church fathers, Augustine in particular, made major strides in taking the sayings and lives of Jesus and the apostles and making of them a morality for functioning societies. Augustine stressed the new commandment of love prescribed by Jesus and promulgated by Paul, but practically he saw the need for rules – he thought the Ten Commandments to be binding on Christians – and reaffirmed the significance of societal laws promoting harmony and social justice. Distinguishing between the earthly city found here in this world and the heavenly city towards which we strive and which will be our reward, he emphasized that the former "seeks an earthly peace, and the end it proposes, in the well-ordered concord of civic obedience and rule, is the combination of men's wills to attain the things which are helpful to this life." In this, there is no conflict with the latter, for "though it has already received the promise of redemption, and the gift of the Spirit as the earnest of it, it makes no scruple to obey the laws of the earthly city, whereby the things necessary for the maintenance of this mortal life are administered; and thus, as this life is common to both cities, so there is a harmony between them in regard to what belongs to it" (Augustine 1972, Book XIX, chapter 17, 326–7).

Particularly influential have been Augustine's thoughts on war, and the conditions under which the Christian might take up arms (in a "just war"). Conflict must be carried on only in the face of an unjust aggressor, under legitimate authority, and under restraint. One cannot simply defend oneself. "I do not approve of killing another man in order to avoid being killed oneself unless one happens to be a soldier or public official and thus acting not on [one's] own behalf but for the sake of others, or for the city in which [one] lives" (Augustine 1983, 148). And mercy must be shown to the vanquished. "Just as we use force on a man as long as he resists and rebels, so, too, we should show him mercy once he has been vanquished or captured, especially when there is no fear of a further disturbance of the peace."

This thinking remains influential. But it fell to Thomas Aquinas to articulate and defend the position that was to become definitive for

Catholic thought: a position deeply indebted to the newly discovered Aristotle and which, being based on observation and reason, could go beyond simple biblical teaching to provide norms for situations quite beyond the ken of Jesus or his immediate followers. Absolutely crucial is Thomas's thinking on the subject of law, distinguishing "eternal law" from "natural law," and these two from "human law." The first, eternal law, refers to God's intentions for the world and the constraints under which He has put it. "Therefore the ruling idea of things which exists in God as the effective sovereign of them all has the nature of law. Then since God's mind does not conceive in time, but has an eternal concept, . . . it follows that this law should be called eternal" (Aquinas 1966, 19–21; *Summa Theologiae* 1a2ae, 91, 1). The second, natural law, is the way in which rational beings (we ourselves) participate in eternal law. It is the working out of eternal law in the context of the human frame and mind. "They join in and make their own the Eternal Reason through which they have their natural aptitudes for their due activity and purpose. Now this sharing of the Eternal Law by intelligent creatures is what we call 'natural law'" (Aquinas 1966, 23; *Summa Theologiae,* 1a2ae, 91, 2). Finally, the third, human law, is what we have and devise, in the light of our need to obey the natural law, which is in turn the eternal law.

Just as from indemonstrable principles that are instinctively recognized the theoretic reason draws the conclusions of the various sciences not imparted by nature but discovered by reasoned effort, so also from natural law precepts as from common and indemonstrable principles the human reason comes down to making more specific arrangements. Now these particular arrangements human reason arrives at are called 'human laws' provided they fulfill the essential condition of law already stated. (Aquinas 1966, 27; *Summa Theologiae* 1a2ae, 91, 3)

Natural law is the key mediating notion in Thomas's ethics. Things, including animals and humans, have certain natural tendencies – ends or goals ("final causes" in the Aristotelian system which structures the discussion) – and their nature displays and reveals these ends. For animals and humans the ends are preservation, life, and (for humans) rational thought and activity. Natural law reflects eternal law and may be encoded in and enforced by human law, but it is itself something discoverable independently through reason and observation: looking at organisms, including humans, seeing the ends that they and their parts serve, and judging accordingly. By example, take homosexual activity. We know that this is immoral because the Bible tells us so (eternal law), and it is something proscribed by human laws. But natural law shows us why it is really wrong, for penises and vaginas were clearly made for heterosexual copulation with the resulting pregnancy and childbirth. To use one's organs in some other way is unnatural: ultimately, of course, an insult to God, Who is creator of all things and Whose creation is entirely good.

The Protestants

In line with their theology, the reformers took a position more directly based on biblical sources. Justification by grace was their theological foundation, something which might be thought to lead to moral complacency. Remember that it is the Pelagian heresy to think that one can buy one's way into the Kingdom of Heaven through good works, and so one might expect the Protestants to believe that all moral effort is worthless. But this is far from true. "Faith by itself, if it has no works, is dead" (James 2:17). One's acts may be worthless compared to one's sins, but the God-touched person acts morally precisely because of this rather than out of a sterile sense of duty, hoping thereby to win praise. Luther compared the Christian to a young lover.

When a man and a woman love and are pleased with each other, and thoroughly believe in their love, who teaches them how they are to behave, what they are to do, leave undone, say, not say, think? Confidence alone teaches them all this and more. They make no difference in works: they do the great, the long, the much, as gladly as the small, the short, the little, and that too with joyful, peaceful, confident hearts. (Luther 1915, 1, 191)

Not that Luther wanted to preach a reformation in society corresponding to his reformation in religion. He warned against revolution and rebellion, taking a particularly dim view of those who rose up against the authorities. The people in power were to him analogous to parents, and in line with the commandment he would have us obey and honour them. "Through civil rulers as through our parents, God gives us food, house and home, protection and security. Therefore since they bear this name and title with all honor as their chief glory, it is our duty to honor and magnify them as the most precious treasure and jewel on earth" (Luther 1959, 29–30). Conversely, those who rebel against authority are wrong and deserving of punishment. "What we seek and deserve, then, is paid back to us in retaliation. . . . Why, do you think, is the world now so full of unfaithfulness, shame, misery, and murder? It is because everyone wishes to be his own master, be free from all authority, care nothing for anyone, and do whatever he pleases" (30).

Calvin likewise stressed the importance of obedience to the authorities and of following the rules of the state. After all, he had been trained as a lawyer! But for him the chief emphasis was on the sovereignty of God: all happens through and because of Him. For this reason, rulers act by His authority, and we find Calvin more ready than most to embrace a democratically elected leadership and to justify rebellion against a false ruler.

And how absurd it would be that in satisfying men you should incur the displeasure of him for whose sake you obey men themselves! The Lord, therefore, is the King of Kings, who, when he has opened his sacred mouth, must alone be heard, before all and above all men; next to him we are subject to those men who are in authority over us, but only in him. If they command anything against him, let it go unesteemed. . . . And that our courage may not grow faint, Paul pricks us with another goad: That we have been redeemed by Christ at so great a price as our redemption cost him, so that we should not enslave ourselves to the wicked desires of men – much less be subject to their impiety. (Calvin 1960, 1520–1, IV, 20, 32)

Not that Calvin had time for those rulers or states which eschewed violence out of a false reading of Scripture. "For it makes no difference whether it be a king or the lowest of the common folk who invades a foreign country in which he has no right, and harries it as an enemy. All such must, equally, be considered as robbers and punished accordingly" (Calvin 1960, 1499, IV, 20, 11). This justifies both a police force within the state and a standing army for defence against attack.

In making this argument about the need for an army and the justified use of force, Calvin was writing against the more radical branches of the Reformation: the Anabaptists. They would have nothing to do with violence and insisted on a literal reading of Christ's commandments, trying often to live communally without private property, according to what they saw as Jesus' direct prescriptions. In this tradition, the Quakers were ardent pacifists, and although they did not withdraw physically and live apart as did others, like the Mennonites, they eschewed the baubles of the world like fine clothes and honours and titles. "It is not lawful to give to men such flattering titles as Your Holiness, Your Majesty, Your Eminency, Your Excellency, Your Grace, Your Lordship, Your Honour, Etc., nor use those flattering words, commonly called COMPLIMENTS" (Barclay 1908, quoted in Beach and Niebuhr 1955, 320).

Even to this day, there is a literalism to the Quaker reading of the Sermon on the Mount which contrasts strongly with their far more relaxed and moderate attitude towards other passages in the Bible, like the early chapters of Genesis. Not that they concerned themselves only with what (today) seem like trivialities and cultural ephemera, such as whether to address one's superiors as "you" rather than "thee" and "thou," and whether to take one's hat off in the presence of the king. Quakers took very seriously the claim by Jesus, reinforced by Paul, that humans are all equal in the sight of God. For them, each person is blessed by the presence of God within the breast, the "inner light." It was a theology like this which led to their being early involved in and always at the forefront of the movement to abolish slavery: a tradition which has continued to this day. (For a modern-day reformed reading of the Gospels, see Murphy 1997.)

One could go on listing further figures and refinements on the views just expounded. There is the calm rationalism of the eighteenth-century Anglicans and the emotionalism of the evangelical movement led by John Wesley: something which reached out beyond the middle classes to the poor and dispossessed, urging them to improve themselves for the greater glory of God. "Having, first, gained all you can, and, secondly, saved all you can, then 'give all you can'" (Wesley n.d., 706). There are the social movements of the nineteenth century, particularly towards its end, when concerned Christians became increasingly concerned with industrialism, recognizing the costs in lives and morality to human dignity. Analogously, there are movements in the twentieth century – for instance, liberation theology, particularly powerful in Catholic circles in South America, which uses insights of Marxism to put the church on the side of the poor against the rich and powerful.

Anything Goes?

By this stage the cynic may be concluding that far from there being such a thing as Christian ethics, there are as many positions as there are writers on the subject. Simply nothing has been barred to those acting in the name of their Lord, and frequently quite contradictory courses of action have been urged as the true Christian way forward. Christians have defended property, Christians have decried property. Christians have defended making war, Christians have been pacifists. Christians have been slaveholders, Christians have been abolitionists. Christians have condemned homosexual behaviour and birth control and abortion, Christians have (especially recently) cherished homosexual behaviour and promoted birth control and defended abortion. From the viewpoint of normative ethics, there is simply no conclusion to be drawn about Christian thinking and behaviour.

At one level, this is surely true. "If you see that there is a lack of hangmen, constables, judges, lords or princes, and you find that you are qualified, you should offer your services and seek the position, that the essential governmental authority may not be despised and become enfeebled or perish" (Luther 1962, 95). If Martin Luther can say this, then just about anything seems to be open. Yet of course, at another level, this is not true. However one decides and acts, one ought to be infused with Christian love: not just towards one's family and friends, but towards one's enemies also. If one convinces onself, for instance, that the Christian way involves physical force – and, to take the paradigm case, many Christians found that very little convincing was needed in the face of Adolf Hitler – then much as one may hate the acts, the intention must always be one of love towards even the vilest of human beings. And this must govern one's own acts. Torturing Hitler might have been very satisfying. It would not have been Christian.

More positively, however far one may think the love commandment extends (more on this later), as a Christian one has an obligation to help the poor and the sick and the homeless: the widow and the orphan and the prisoner and the dying and destitute. Mother Teresa's Christianity and her activities in Calcutta were not coincidentally linked. Moreover, even though one may perhaps differ with her over some of her views – the impermissibility of birth control and abortion, for instance – one can understand the Christian nature of her emphasis on personal restraint, particularly in sexual matters. Christians go (and have gone) all of the way from rigid denial of all sexual activity except for limited sexual acts within marriage (itself judged less than the most desirable state), to very tolerant acceptance of virtually all of the ways in which humans seek erotic satisfaction. There is nevertheless a presumption in favour of self-discipline (Ruse 1988b). To pretend otherwise is to ignore the Christian heritage.

Foundations

What now about the question of foundations? What is Christian meta-ethics? Most obviously and most centrally, the Christian puts his or her faith in God as revealed to us through Jesus Christ – in his love and care for us – and finds the justification for substantive ethics in God's will. We should do that which He wants. And why should we do that which He wants? Well, ultimately because that is what He wants us to do: end of argument! Ours is not to reason why. This is a tradition which goes back to Jewish thought. God tells Abraham to sacrifice his son Isaac: a son born only after much trial. There is simply no question in Abraham's mind that this is what he must do. God has spoken and issued His orders, and that is an end to matters. The same is true for the Christian. In the Sermon on the Mount, Jesus does not reason with his audience, trying to persuade them to his opinion. Rather, he lays on the line the expectations of the Christian. These are the things you must do, because these are the things that God wants. And the same holds for the Christian today. The way of the Cross is a demand that God puts on His followers.

In a sense, this all sounds like a deal or a bribe. You do what I ask, and I will offer you goodies in return: a land flowing with milk and honey perhaps, or relief from your sins, or eternal life. Certainly, this is rather the way that Abraham's covenant with Yahweh comes across. You keep your side of the bargain, and I will keep mine. The mark of those keeping the covenant is not presented as anything other than something the Lord has decided on as a sign. "This is my covenant, which you shall keep between me and you, and your descendants after you: Every male among you shall be circumcised." (Gen. 17:10) If the Lord had decided on a facial tattoo, like those worn by some African tribes, that would have done as well and have had as much justification.

However, there has to be more than this for the Jews, and certainly for the Christians. Plato, in the *Euthryphro*, four hundred years before Jesus, put his finger on crude versions of the divine command theory. Is that which is good, good simply because God commands it, or does God command it because it is good? Could God simply command the arbitrary on everything? Could God, for instance, simply make rape an ethically acceptable practice? Ethically mandatory, indeed? Surely not! In which case, it would seem that God commands things because they are good, which seems to make the good independent of God's will or intention. From a metaethical perspective, His will is irrelevant, although obviously He backs up morality with the divine carrot and stick.

Of course, things are not quite this simple, either way (Quinn 1978). God is creator of everything, so ultimately morality has to rest in His will and His creative power. Yet while He is omnipotent, we have seen that this does not imply that He can do the impossible. Moreover, He is in His nature infinitely good. He could do or wish ill, but it is not of His nature to do this. Unlike us, God is not tainted with original sin. Hence, on the one hand, God is constrained by practical necessity. To borrow the point that Kant made – since Kant was the child of deeply Pietistic (Anabaptist) parents, it was hardly a surprise that he made it – God could not make a functioning and happy society if everyone could lie and cheat and break promises with impunity. This is simply not possible, and God has never claimed the ability to do the impossible. On the other hand, God could only want what is best for us, and He could only make and endorse the rules which serve us best. If God could make a society which functions like the society under the Nazis – some people do well, but others suffer terribly – He would never do so. He might be free to do so, but He would not do so. I am free to go with an axe in the night and murder my children, but I would no more do so than God would do something similar.

For the Christian there is a necessity to morality – a universality – which is endorsed, demanded, by God, which is not capricious or arbitrary. God wants what is right, God wills what is right, God demands what is right, and through His grace forgives us when we fall short. But God does not simply make it all up as He goes along. Morality is part of the nature of things. The Catholic doctrine of natural law brings this out most fully – God having made male and female, certain sexual acts are by necessity natural and proper – but it is a general conclusion held by all Christians. If you are going to argue that something is morally acceptable, then you must show that it is natural. Consider, for example, debates about birth control and population restraint. Back when most humans died before themselves reproducing – because of childhood illnesses and the like – having as many children as possible was entirely natural. Now today, those who argue for birth control base their case on the existence of modern medical practices and so forth, which preserve people until adulthood and thus indirectly contribute to a horrendous population explosion. It is argued therefore that it is no longer natural – and hence no longer mandated for the Christian – to have as many children as possible. One must practice restraint or protection or some such thing.

Likewise, but from the opposite pole, those who oppose birth control argue that this practice is not natural: "Since . . . the conjugal act is destined primarily by nature for the begetting of children, those who in exercising it deliberately frustrate its natural power and purpose sin against nature and commit a deed which is shameful and intrinsically vicious" (Pius XI 1933, 25). But note how the Catholic Church today will allow the rhythm method of control, which is judged natural. Note also how, in the West, most Catholics ignore the church's teaching on the subject of contraception. They feel that the Church is out of step with what is now right and proper and natural. Apart from anything else, people today recognize that sexual intercourse in humans promotes pair bonds – women are continuously receptive and do not wait to come into heat – and these bonds are important for the care of human infants, offspring which have such a slow and demanding process of development. Contraception can therefore be defended on strictly Thomistic grounds.

Leave now the details and examples. We now have before us a sketch of the Christian position on morality. Let us turn next to the traditional position on evolution and ethics.

CHAPTER TEN

Social Darwinism

For all that traditional evolutionary ethics is known as social Darwinism, its greatest debt is to Darwin's fellow English evolutionist Herbert Spencer. It is true that, in the *Descent of Man,* Darwin makes moral judgements and prescriptions – of an entirely conventional upper-middle-class Victorian ilk. It is true also that Darwin's great authority as a scientist was important. Even those who had never read the *Origin* often used Darwin and his standing to bolster conclusions. But it was Spencer who really counted, and thus it is to him that we turn first.

Herbert Spencer

Prima facie, at the substantive level, the social Darwinian strategy is simple and direct. One ferrets out the nature of the evolutionary process – the mechanism or cause of evolution – and then one transfers it to the human realm, arguing that that which holds as a matter of fact among organisms holds as a matter of obligation among humans (Ruse 1986a). Herbert Spencer would seem to epitomize this strategy. He started with the struggle for existence and the consequent selective effects: a connection which he made in print in 1852, years after Darwin had made the connection but years before Darwin published. He then transferred this to the human realm: not much to do here, actually, since Spencer speculated on selective effects showing themselves in the different natures and behaviours of the Irish and the Scots. He concluded that struggle and selection in society translates into extreme laissez faire in socioeconomics. The state should stay out of the way of people pursuing their own self-interests and should not at all attempt to regulate practices or redress imbalances or unfairness. Libertarian license, therefore, is not only the way things are but the way they should be (Spencer 1852a,b, 1862, 1892).

Now there is certainly some truth in this picture. So far was Spencer committed to laissez faire that he would not even have had the state provide lighthouses! Let ship owners band together and provide them if they have need of them. Again and again he fulminated against any kind of state aid for the poor, and he was scathing about those who disagreed.

Blind to the fact that under the natural order of things, society is constantly excreting its unhealthy, imbecile, slow, vacillating, faithless members, these unthinking, though well-meaning, men advocate an interference which not only stops the purifying process but even increases the vitiation – absolutely encourages the multiplication of the reckless and incompetent by offering them an unfailing provision, and *discourages* the multiplication of the competent and provident by heightening the prospective difficulty of maintaining a family. (Spencer 1851, 323–4)

And one can find similar sentiments in the writings of Spencer's followers, especially those of the turn-of-the-century American sociologist William Graham Sumner: "The facts of human life . . . are in many respects hard and stern. It is by strenuous exertion only that each one of us can sustain himself against the destructive forces and the ever recurring needs of life; and the higher the degree to which we seek to carry our development the greater is the proportionate cost of every step" (Sumner 1914, 30).

But there is – there has to be – much more to the story than this. For all that he made the independent discovery of natural selection, it was always a very minor part of the Spencerian evolutionary picture. The Lamarckian inheritance of acquired characteristics was the chief driving force of Spencer's evolutionism (Spencer 1864). And, even more anomalous, many of Spencer's harshest-sounding pro laissez faire claims occur in his *Social Statics,* written before he became a full-fledged evolutionist! The fact is that, if anything, his ethical theory can be attributed to his background of Protestant nonconformism, which saw the poor laws and the like as keeping much of the population in a state of perpetual poverty and dependency, thus serving the ends of the rich and powerful (represented by the Anglican church) who inherit their wealth and status and who have no fear of the threat of competition from the more gifted or industrious (Spencer 1904; Duncan 1908). Spencer's evolutionism certainly moved in to confirm and support these views, but there was no simple deduction of ethics from biology.

Morever, Spencer was far from denying the worth of any individual charity. It was rather state-supported institutions of charity which he opposed. The same is very much true of his followers. John D. Rockefeller, the founder of Standard Oil and one of the notorious businessmen at the beginning of this century, was openly in favour of denying state interference. He spent much of his time opposing the federal government as it strove to break up the monopoly he had established over the distribution and sale of fuel oil. He justified himself in Darwinian terms, saying that the fit do and should survive. Yet from the beginning he had tithed himself, and he always gave generously to charity. Likewise, Spencer enthusiast Andrew Carnegie, founder of United States Steel, claimed that no rich man should die rich. He gave much to the founding of public libraries: places where poor but gifted children could go and thereby improve themselves and raise themselves up in society (Russett 1976; Bannister 1979).

Alternatives to "Laissez Faire"

Like those of Spencer himself, the motivations of his followers owed much to their Christian heritage: often to a Presbyterian childhood, stressing thrift, hard work, the virtues of industry and business, and (above all) God's having judged some of us as sheep and others as goats. Not that the evolutionism was a veneer. It was deeply felt, even if it was not the only factor. The same is true also of evolutionary ethicists who wanted to draw conclusions very different from those of Spencer: the people who wanted to argue that far from supporting capitalism and denying socialism, evolution promotes a strong state and the attempt to diminish inequality through its controls. Notwithstanding his close friendship with Spencer, Thomas Henry Huxley was a paradigm here. He was one of a group of scientists, increasingly successful at finding themselves and their students positions of authority and power within universities and the civil service, who grew correspondingly sceptical of extreme laissez faire. They saw the virtues of a bureaucracy and of state intervention, in education and elsewhere (Huxley 1871).

Some were more extreme. Alfred Russel Wallace, the codiscoverer of natural selection, was ever an ardent socialist who thought that the state can and should regulate people's lives for the better (Wallace 1905; Marchant 1916; Ruse 1996a). He was motivated in his thinking by the early influence of the Scottish mill owner and socialist Robert Owen, and he drew comfort from a deep enthusiasm for spiritualism – seeing a guiding force presiding over human fortunes. But Wallace's justification was evolutionary. Against Darwin, he thought that selection favours groups as well as individuals, and he concluded that a state founded and run on socialist principles would be better – and better prepared for the future – than one which simply bowed before market forces. (See also Jones 1980.) Similar sorts of reasoning led the Russian Prince Petr Kropotkin to his anarchism. He believed that there is a natural sympathy existing between people (and animals), which he called "mutual aid." To Kropotkin, coming from nineteenth-century Russia – a vast preindustrial society where the chief threat to life lay in the elements – it seemed self-evident that evolution must work for good and sympathy rather than harm and competition. "The animal species, in which individual struggle has been reduced to its narrowest limits, and the practice of mutual aid has attained the greatest development, are invariably the most numerous, the most prosperous, and the most open to further progress. . . . The unsociable species, on the contrary, are doomed to decay" (Kropotkin 1902, 293).

War and Peace

But this – warm and cuddly though these sentiments may be – surely ignores the main thrust of social Darwinism, that which deservedly has given it a bad name. Darwin's ideas were taken up on the continent, in Germany in particular, leading to warlike sentiments as struggle between nations was justified as natural in the name of Darwin. They provided a rationale for the militarism which led to the First World War as well as a theoretical justification for the vile philosophies which followed in its wake: communism and national socialism. "He who wants to live must fight, and he who does not want to fight in this world where eternal struggle is the law of life has no right to exist" (Hitler 1939, 242). Thus Adolf Hitler in *Mein Kampf*.

Even here, however, matters are mixed, with respect both to influences and to consequences. Start with militarism. One can certainly find force and expansion justified in the name of Darwin. Huxley was one who thought that force and violence tend to be natural, although he was explicit that this is not moral but rather to be combatted in the name of morality. Others went farther down the savage bloodstained-ape path, recognizing our nature as the way things are and are destined to be. One writer, in a passage which admittedly perhaps owes as much to Hegel as it does to Darwin, claimed that war is "a phase in the life-effort of the State towards completer self-realization, a phase of the eternal nisus, the perpetual omnipresent strife of all beings towards self-fulfilment" (Crook 1994, 137). Even though not this enthusiastic, others saw it as "more or less normal for men at times to plunge back down the evolutionary ladder . . . to break away from the complex conventions and routine of civilized life and revert to that of the troglodites in the trenches." And again, "Man has always been a fighter and his passion to kill animals . . . and inferior races . . . is the same thing which perhaps in the dark past so effectively destroyed the missing link between the great fossil apes of the tertiary and the lowest men of the Neanderthal type. All these illustrate an instinct which we cannot eradicate or suppress, but can best only hope to sublimate" (143–4).

Some, however, stressed that war is only a temporary primitive phase and urged peace in the name of evolution. Spencer was in the forefront of this group. Again, perhaps, nonconformism was significant – there were Quaker elements in his past – but his justification was evolutionary also. Indeed, it was the essence of Spencer's position that Lamarckian strife means a rise up the chain of being with a consequent decline in fertility (as effort goes into brainpower rather than reproduction), with a falling away of the struggle and universal peace. This fit in with his belief that what is best for business and unrestrained laissez faire is free trade and open frontiers. The last thing Spencer wanted was the waste of military spending and the erection of barriers to unrestricted commercial intercourse.

Others felt the same way down to and through the First World War (Mitman 1990). But what of the ideologies? Start with communism. Both Engels and Marx liked the *Origin* (although thinking it crudely empirical and English), and, at the grave of Marx, Engels went so far as to say that Marx did in the social world what Darwin did in the biological world (Young 1985). There was therefore always a warm place for Darwinism in the Soviet system, although one should note that much of the thinking was more Germanic than Darwinian: Engels's posthumously published *Dialectics of Nature* shows strong Hegelian-*Naturphilosoph* influences. When it came to practical agriculture, the sorry story of Lysenko shows that the Darwin-complementing Mendelism counted for little (Joravsky 1970). Lamarckian change through effort had a natural appeal that Darwinian change through selection did not. One can therefore hardly say that Darwinism really influenced Marxism or its consequent communism: it was mainly used as justification. Scholarship shows, however, that particularly in America many who called themselves Marxists owed more to evolutionism – to Spencer particularly – than they did to Marx (Pittenger 1993). To this often was added a dash of the biogenetic law – that ontogeny recapitulates phylogeny – as it was argued that societies go from primitive (and homogeneous) up to the most developed (and heterogeneous), which are communistic (Richards 1987).

National socialism likewise bears an ambiguous relationship to evolutionary ideas. Ernst Haeckel – for all that he claimed to be Darwin's German disciple – favoured more of a group perspective than an individual-based one. He used this to argue the virtues of the integrated state with a strong military, efficient civil service, and well-supported universities: Prussia under Bismarck, in fact! One can see here the outlines of the Third Reich, and to this one can add other elements: a cherishing of the highest form of human, who just so happens to be Nordic and who excludes other races, including blacks and Jews. There were other ideas which were to resurface in the 1930s, including views about the virtues of selective breeding and the elimination of the unfit (eugenics), as well as a strong opposition to Christianity as the religion of the weak. Hitler notoriously was contemptuous of Christianity, arguing that it tried to put humankind outside of or beyond nature when we must realize that we are all part of the living flow of nature. He even condemned Christianity for its opposition to evolution! (Gasman 1971, 168, 113n)

But this is only a very small part of the story, and the overlaps frequently owe more to common cause than to cause and event (Kelly 1981; although see Gasman 1998). Hitler's philosophy owed most to the Volkish ideology of the nineteenth century, which saw Germans uniquely as the supreme race, threatened by outsiders: threatened above all by the Jew. This led to what has been called "redemptive" or "apocalyptic" anti-

Semitism: an anti-Semitism which has a kind of ontological or religious status (Friedlander 1997). Haeckel, by contrast, may not have much cared for Jews, but his solution was to assimilate them. This was the farthest thought from Hitler's mind. More that this, evolution – as most of the Nazis saw quite clearly – was fundamentally opposed to the ideology of National Socialism. Within evolutionary theory, there is no warrant for saying Germans are uniquely the superior race, there is denial that this can be a permanent state of affairs, there is the connection of all peoples including Aryans and Jews, there is the simian origin, and much more. It is hardly surprising that Nazi celebrations of Haeckel's centenary were muted in the extreme and that evolutionary works were among those proscribed by the party. The Nazis knew who were their friends and who were not.

Women

What about social ideals? One place where social Darwinism is often faulted – where Darwinism generally is faulted – is with respect to females. "The support offered by a science that was for the most part accepted in its day as objective and value-free immeasurably strengthened patriarchy for decades to come. The *Origin* provided a mechanism for converting culturally entrenched ideas of female hierarchy into permanent, biologically determined, sexual hierarchy" (Erskine 1995, 118). There is truth in this assessment. Women are described in demeaning and belittling ways, and then this is used as a justification for moral attitudes and social policy. Certainly, if you look at the *Descent of Man*, you will find some pretty conventional Victorian sentiments about the sexes and their relative worth. "Man is more courageous, pugnacious, and energetic than woman, and has a more inventive genius" (Darwin 1871, 2, 316). In compensation, woman has "greater tenderness and less selfishness" (326). You could be reading a novel by Charles Dickens.

Even those relatively favourable to women, like Huxley, usually ended up by concluding regretfully that females are inferior to males and that this justifies discriminating actions, such as excluding women from societies and universities. And if sexual selection is not at work – Darwin thought that the differences between men and women are a direct function of this mechanism – then the biogenetic law steps in.

In the animal and vegetable kingdoms we find this invariable law – rapidity of growth inversely proportionate to the degree of perfection at maturity. The higher the animal or plant in the scale of being, the more slowly does it reach its utmost capacity of development. Girls are physically and mentally more precocious than boys. The human female arrives sooner than the male at maturity, and furnishes one of the strongest arguments against the alleged equality of the sexes. The quicker appreciation of girls is the instinct, or intuitive faculty in operation; while the slower boy is an example of the latent reasoning power not yet developed. Compare them in after-life, when the boy has become a young man full of intelligence, and the girl has been educated into a young lady reading novels, working crochet, and going into hysterics at the sight of a mouse or a spider. (Allan 1869, cxcvii)

Again, however, as with war and violence, we find that others have claimed precisely the opposite, also in the name of evolution! Just as Darwinism has been used as a vehicle for sexist views, it has also been used as a vehicle for feminist views! Consider Wallace. He held the view that human progress depends ultimately on female sexual selection. Men cannot be trusted with the future of our race. Fortunately, in days to come, young women will take over the reins, choosing as mates only those males with the highest moral and intellectual properties. Thus upward progress is guaranteed.

In such a reformed society the vicious man, the man of degraded taste or of feeble intellect, will have little chance of finding a wife, and his bad qualities will die out with himself. The most perfect and beautiful in body and mind will, on the other hand, be most sought and therefore be most likely to marry early, the less highly endowed later, and the least gifted in any way the latest of all, and this will be the case with both sexes. (Wallace 1900, 2, 507)

You will get a kind of natural eugenics, and this will lead to general improvement of the human species: "This cause continuing at work for successive generations will at length bring the average man to be the equal of those who are among the more advanced of the race." You may object that if Wallace truly thought any of this to be remotely possible – that young women in the future will freely mate with only the best of the male crop – then his knowledge of human nature was about on a par with his touching beliefs in the integrity of spiritualists. But the fact remains that Wallace made his claims in the name of evolution. And, after Darwin, if any nineteenth-century evolutionist deserves the label "Darwinian," it is he.

Today, consider the biological anthropologist Sarah Blaffer Hrdy. In her *The Woman that Never Evolved* (1981), she has put forward a feminist picture of human evolution and resultant nature that would not be out of place in *Ms.* magazine. She argues that the reason why human females uniquely conceal their ovulation is to keep the poor males on a parental-care-giving string. Because, from one brief sexual encounter, they do not know, they cannot know, whether or not they fathered the resultant children, they must stay around: keeping out rivals but incidentally also helping with the child rearing. If they do not, the women cut them off. Lysistrata in the genes, one might say. Feminism made flesh, one might also say. And, like Wallace, Hrdy cannot be dismissed as a side phenomenon, a kind of scientific freak. She is one of the more successful evolutionists active today – a member of the National Academy of Sciences, no less.

Biodiversity

With Hrdy we have moved right out from the nineteenth century up to the end of the twentieth. One could spend much time on the prescriptions of social Darwinians in our century. There is Julian Huxley, the biologist grandson of Thomas Henry Huxley, who spent much time promoting the virtues of science. Ever an evolutionary ethicist, it was he who (as first director general of the organization) insisted on the Science being added to UNESCO (Huxley 1948). Then there was G. G. Simpson, the scourge of extraterrestrials, who promoted the American political system in the name of evolution. "Democracy is wrong in many of its current aspects and under some current definitions, but democracy is the only political ideology which can be made to embrace an ethically good society by the standards of ethics here maintained" (Simpson 1964, 321). But, rushing past these and others, let me turn again to Edward O. Wilson. One would expect that he too would draw moral implications from his evolutionary biology, and he does not disappoint:

Camus said that the only serious philosophical question is suicide. That is wrong even in the strict sense intended. The biologist, who is concerned with questions of physiology and evolutionary history, realizes that self-knowledge is constrained and shaped by the emotional control centers in the hypothalamus and limbic systems of the brain. These centers flood our consciousness with all the emotions – hate, love, guilt, fear, and others – that are consulted by ethical philosophers who wish to intuit the standards of good and evil. What, we are then compelled to ask, made the hypothalamus and limbic system? They evolved by natural selection. That simple biological statement must be pursued to explain ethics and ethical philosophers, if not epistemology and epistemologists, at all depths. (Wilson 1975, 3)

What exactly does this all mean in practice? Fundamental to Wilson's thinking is his belief that a biologically inherited love of nature is possessed by all humans. Apparently we humans have evolved to such a point that we have a symbiotic relationship with nature. Without it we would wither and die. We humans not only need nature physically to survive, we need it spiritually. A world of plastic would be deadly, literally. This belief leads easily into ardent public enthusiasm for "biodiversity": the life on earth, its complexity, and its interrelations. For Wilson, the key to the organic world is its heterogeneity, its fascinating and ever-changing richness of forms and groups and relationships and antagonisms and dependencies and so much more. There are millions and millions of species, and each one is entwined with hundreds of others: as neighbours, as competitors, as predators, as prey, as parasites, as hosts, as destroyers, as helpers, as so many other things. And we humans are right in the heart of it all, using and depending on the biology that surrounds us and of which we are a part.

The living world system is under continual threat from natural disasters and the like. But fortunately it has the ability to spring back. The terrible explosion on the island of Krakatau in 1883 was followed by recolonization of what was left. "Today you can sail close by the islands without guessing their violent history, unless Anak Krakatau happens to be smoldering that day" (Wilson 1992, 23). Not even the great global extinctions of the past were enough to destroy biodiversity forever. Nonetheless, the time needed for recovery was extensive. Major extinctions required 25 million years or more (31). Now, alas, we face the biggest extinction of them all: the human-caused extinction. We are destroying species at a phenomenal rate, and most tragically among the worst-affected places are the rain forests and jungles of the tropics – Brazil, for instance. We must do something before it is too late. Else we will never see biodiversity again in our lifetimes – or in our children's lifetimes.

This is more than a pragmatic call. It is a spiritual warning. "Only in the last moment of human history has the delusion arisen that people can flourish apart from the rest of the living world." Would that we were more

like preliterate folk, who may not have understood the underlying principles, but who did grasp that "the right responses gave life and fulfilment,
the wrong ones sickness, hunger, and death" (348–9). Humans must heed
the call and respond. Armageddon is here and now, and the fight is on.

Comparisons

Pause for a moment and compare the evolutionists' claims and demands
to Christian moral prescriptions. The impression that most people have –
an impression often reinforced by partisans on both sides – is that, ethically, Christianity and social Darwinism could not be farther apart. The
former stresses peace and love and forgiveness. The latter, introduced
precisely because Christianity simply did not apply in a modern urban
industrialized society, stresses competition and struggle and success and
failure. The Christian believes it is easier for a camel to pass through the
eye of a needle than for a rich man to enter the Kingdom of Heaven; that
one should succour the widow and orphan; and that one should turn the
other cheek. The social Darwinian thinks that this is unforgivably tender-
hearted, with no possible practical implications. Wealth is to be cherished, power also; widows, orphans, and losers generally must suffer in
life's stern struggles; and war is a necessary (and often beneficial) part of
existence. Gordon Gekko, in the movie *Wall Street*, is the paradigmatic
social Darwinian.

Already we have sensed that this is about as far from the truth as it is
possible to be. Far from social Darwinism being something formulated in
explicit opposition to and independence of Christianity, often it owes as
much, if not more, precisely to that religion in which most evolutionists
grew to moral maturity. Spencer himself formed his moral system from
elements of his Christian training and culture, well before he became an
evolutionist. This showed right up to the end of his life, when he cam-
paigned most vigorously against militarism. Some, like Rockefeller, were
explicit in their attempts to blend their Christian beliefs and their evolutionary beliefs into one moral synthesis. And even when people went
beyond or rejected Christianity, the influences show through. Wilson's
urge to biodiversity owes at least as much to a Christian appreciation of
the glory of God's creation as it does to anything in Charles Darwin.
Wilson's attitude towards animals and plants parallels precisely the way

today's Christians read the second chapter of Genesis – as God pressing on Adam stewardship over the living world.

Of course, if you want to set up the Christian against the social Darwinian, you can do so. You set the Christian pacifist against the neo-Haeckelian warmonger, or the Christian socialist against the free-market-loving Spencerian, or the Christian feminist against Charles Darwin. But because Christianity and social Darwinism have such broad ranges of interpretation (at the level of substantival ethics), you can do pretty much the opposite also! You can set the Christian warmonger against the Spencerian pacificist, the Christian capitalist or free-marketer against the socialist Wallace or the communist Engels or the anarchist Kropotkin, or Saint Paul against a host of Darwinian feminists, from Alfred Russel Wallace to Sarah Hrdy. Think of your moral cause, and the chances are that you will find Christians who have supported it in the name of their Lord, and Christians who have opposed it in the name of that same Lord. Likewise with traditional evolutionary ethics.

One does see certain tendencies from the Christian position. One sees the same tendencies from Darwinians. No one seems to like lawyers very much. "Woe unto you lawyers also! for you load men with burdens hard to bear, and you yourself do not touch the burdens with one of your fingers" (Luke 11:46). Social Darwinians are careful always to exclude lawyers from their utopias. More seriously, Christians tend to come down on the side of peace and forgiveness. For all that many seem to have the impression otherwise, the same is true of social Darwinians. There is more written showing the futility of warfare from an evolutionary perspective than there is urging men to war and violence (Crook 1994). Either that, or arguing – as did Thomas Henry Huxley in his famous last essay ("Evolution and Ethics") – that evolution can and must be conquered and that we should quell our killer instincts for the sake of peace and harmony.

The overall point is that Darwinians can be Christians (with respect to substantive morality) and that many have been. And those social Darwinians who eschewed religion were often subscribing to precisely the same beliefs as were Christians.

Progress as Justification

Move now to the question of foundations. The social Darwinian's argument is simple and direct. He or she identifies what is believed to be the main causal force of evolution. Then the moral norm – the substantive ethics – is drawn from this cause. It is claimed that this is justification enough. This is the way that evolution acts. This is the way that we ought to act – either promoting evolution as it has been, or (at the very least) preventing evolution from slipping back and destroying that which it has achieved. There is a straightforward move from the empirical – the way that things are, matters of fact, "is" statements – to the moral – the way that things ought to be, matters of obligation, "ought" statements. There is a move from description to prescription.

"Simple and direct." In the opinion of most moral philosophers in the last century, more simplistic than simple, and more misdirected than direct. Famously, the English philosopher G. E. Moore (1903) found Spencerian metaethics to be a prime exemplification of what he labelled the naturalistic fallacy. At a more general level, as many have pointed out, the transgression of the "is/ought" barrier is always illicit. David Hume knew the score here:

In every system of morality, which I have hitherto met with, I have always remark'd, that the author proceeds for some time in the ordinary way of reasoning, and establishes the being of a God, or makes observations concerning human affairs; when of a sudden I am surpriz'd to find, that instead of the usual copulations of propositions, *is,* and *is not,* I meet with no proposition that is not connected with an *ought,* or an *ought not.* This change is imperceptible; but is, however, of the last consequence. For as this *ought,* or *ought not,* expresses some new relation or affirmation, 'tis necessary that it shou'd be observ'd and explain'd; and at the same time that a reason should be given, for what seems altogether inconceivable, how this new relation can be a deduction from others, which are entirely different from it. (Hume 1978, 87)

Later I shall make a virtue of the is/ought distinction. Now, I note only that traditional evolutionary ethicists tend to be supremely untroubled by charges of fallacious reasoning. They are even inclined to agree that generally the move from "is" to "ought" is fallacious: save only in this one case! When one is dealing with evolution, they argue, it is uniquely legitimate to go from the way that things are and have been, to the way that they ought to be.

Why is there this confidence? Is there a missing premise? There most certainly is. To a person, traditional evolutionary ethicists – social Darwinians – believe that the course of evolution is no random meaningless walk (Ruse 1986a). Our old friend, progress, rears its head. Traditionalists believe that evolution is progressive, with humankind at the top. For them, it is therefore our moral obligation, given that progress implies value (what else would it imply?), to cherish and aid and repair, or (at a minimum) not to hinder, the course of evolution. Again and again one finds that this is the move taken. Herbert Spencer (1857) was fanatical about evolutionary progress, from the homogenous to the heterogeneous. So also was Haeckel (1866, 1868) (Figure 9). Fisher (1930) thought that his fundamental theorem guaranteed progress. Julian Huxley (1942) believed that somehow we move up the scale pulling the ladder behind us, so that no other animal can scale the human heights. Simpson (1949) thought that all notions of progress are subjective but came out strongly for the case that there is a progressive evolutionary rise up to humankind. And Edward O. Wilson (1992, 1994) preaches upward progress with the fervour and fanaticism of the revivalist preachers of his youth. One and all believed or believes that evolution itself gives meaning and confers value. Evolution leads to good and to things of great value. Hence, it is the source of our moral obligations.

Progress is an old friend which raises old worries. Some evolutionists, some Darwinian evolutionists, are happy with biological progress. Others are not. Spencer championed it. Thomas Henry Huxley (1893b) came eventually to question it, at the end of his life denying that that which has evolved is necessarily good and arguing that right conduct often involves fighting that which has evolved. Richard Dawkins likes it. Stephen Jay Gould does not. But how one should decide on the issue of progress is not really the question here. It is enough to recognize that social Darwinians appeal in justification of their substantival ethical claims to the supposedly progressive nature of the evolutionary process and of the finished product. That is why moral claims are justified.

Which brings us back to our title question: Can a Darwinian be a Christian? We can leave now those evolutionists who are not Darwinians, whether or not they be progressionists. They are not our concern. We can leave also those Darwinians who are not progressionists: in fact, I shall be speaking more of them in the next chapter. At issue are those Darwinians who are (biological) progressionists and who use their progressionism as justification for their normative moral prescriptions. My sense is that, although of course they do not have to be Christians, if Christianity is their commitment, then this aspect of their Darwinism fits nicely with their religion. Indeed, I would go further than this. My suspicion is that

PEDIGREE OF MAN.

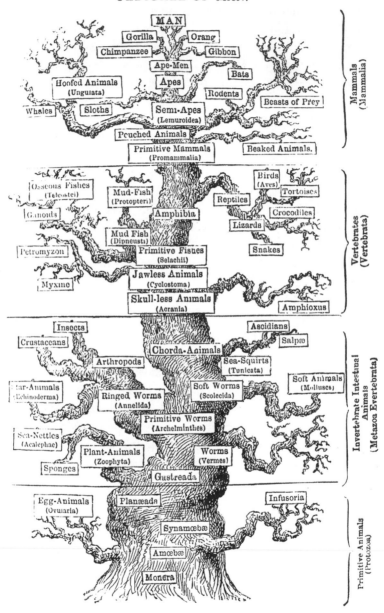

9. The tree of life as drawn by Ernst Haeckel, showing unambiguous progress. (Haeckel, E. *The Evolution of Man*, 1896.)

the social Darwinian who would be a Christian is in a better position than his or her nonbelieving social Darwinian fellows. This holds true on both empirical and theological grounds.

The Christian has already accepted that humans appeared according to God's will and that God lays morality upon us, intending that we should obey. Moreover, Christian morality is natural, in the deep sense that it stems from the way that we are. The female preying mantis may eat her husband; such an action could never be natural or moral for us humans. So, on the one hand, the evolutionary scenario shows nicely why, even if God wants us to do good, what we ought to do is not simply a function of His unguided whim. He must intend for us that which is within the limits of our evolved nature. God could not want us never to think about sex at all, because that is part of the way we have evolved – even though God could want us to think differently about sex than we do. On the other hand, the Christian backing counters all of the worries one might have about illicit bridging of the is/ought gap. The progress of evolution was part of God's plan, God's intention. It is not just a matter of fact, but a matter of divine desire. And as such, it gains moral value. Hence, if as a Wilsonian one promotes moral diversity, one is doing what God wants, for one is thereby looking to the interests of His special creatures. This in itself is all the moral justification one could ever have or want.

I conclude that, far from Christianity's posing problems for the social Darwinian, it plugs some significant gaps in ways not open to the non-believer. You will see in the next chapter that traditional social Darwinism is not the only way to meld Darwinism and ethics. But if you do want to go this traditional way, then Christianity meshes nicely. Sir Ronald Fisher's world system is a paradigm. God's task was to create humans, which He did through His progress-guaranteeing fundamental theorem of natural selection. Selection pushes populations to ever higher points of fitness, thus countering the degenerative effects of the Second Law of Thermodynamics. The human task is to keep humans up and beyond their natural peak. Here the key is eugenical intervention in human breeding patterns. Thus one counters the degenerative effect of the fact that successful humans are precisely those with the smallest families. Biological progress stands behind our ultimate moral obligations, and behind this stands God.

You may not much care for these particular details, but as was so often the case, Fisher got the main picture right.

CHAPTER ELEVEN

Sociobiology

For Edward O. Wilson, as for Herbert Spencer and Julian Huxley before him, Darwinism is a substitute for Christianity: a secular religion for a new age. By now you should realize that you do not have to read Darwinism in this way – most professional evolutionists today cringe rather at this kind of activity – and that, even if you do, you are probably singing the same good old songs that have sustained Christians down through the ages. Indeed, you probably first learnt the songs at Sunday school in your childhood! No one could doubt the authenticity of Wilson's deeply religious nature or the power of his burning moral vision, but his arguments purportedly showing Darwinism and Christianity to be mutually exclusive are simply not well taken. Indeed, if you insist on making a religion of your science, then your best strategy might be to join forces with Christianity rather than trying to set up your own church.

Is this then the end of matters? If you would prefer not to make a religion of your science, has Darwinism nothing more to say? Can modern evolutionary theory tell us nothing about morality, at either the substantive or metaethical level? A totally negative answer to these questions would be surprising, if only because the past thirty years have seen major advances in the Darwinian understanding of the evolution of social behaviour. That area where morality most comes into play, the interactions between individuals in a cooperative or social manner, has been the subject of intense scrutiny by Darwinians, who think that they have completely transformed our thinking on the question. It is the development of the science and the implications to be drawn which are the subjects of this chapter.

The Evolution of Social Behaviour

Charles Darwin himself always recognized that behaviour is important. It is no good being strong and handsome if you do not have the desire and ability. He recognized also that while all behaviours are significant, some are more difficult to explain than others. If a sheep flees a wolf, this is behaviour by both predator and prey, and nothing very surprising. If an organism does something for some other organism, then this does call for special attention. If indeed life is a bloody struggle for survival and reproduction, why would an organism behave "altruistically" rather than "selfishly," caring only for itself? Epitomizing the puzzle, why do the social insects evolve as they do, with sterile workers devoting their whole lives to the good of the nests within which they live? Why not look after number one exclusively?

There is a simple answer to the evolution of altruism in general and of the social insects in particular. It is just a question of selection favouring the group – the ant nest, for instance – rather than the individual. A simple answer, but inadequate. Darwin realized that any adaptations favouring the group at the expense of the individual will prove highly unstable. They will always be at risk of crumbling under an individual-favouring alternative. Over the long run an adaptation might revert to individual benefit via the group; unfortunately, in the short term the individual who takes advantage of the efforts of others while not returning in kind will be at the greatest advantage (Ruse 1980, 1989).

It was the 1960s before evolutionists found ways to tackle Darwin's problems about sociality. A major breakthrough came thanks to the then–graduate student William Hamilton (1964a,b), who pointed out that the best-known social insects, the hymenoptera (the ants, the bees, and the wasps), have a haploid-diploid reproductive system. Males have only mothers and only a half set of chromosomes, whereas females have both mothers and fathers and a complete set of chromosomes. Hamilton was able to show how workers nevertheless serve their own biological interests: it is more profitable from an evolutionary perspective to raise fertile sisters than fertile daughters (Figure 10). He was able to show also that his model, which is now covered by the generic term "kin selection," can be extended to other organisms. And indeed, his work is still today considered the paradigm of an explanation of biological "altruism": features

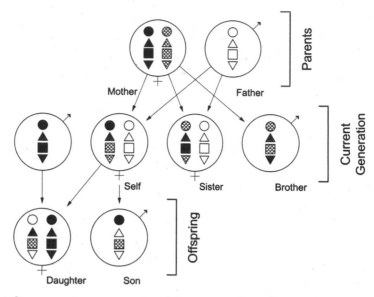

10. A diagrammatic representation of the genetic relationships in the *Hymenoptera*. Females are diploid (that is, have two half-sets of chromosomes); males are haploid (that is, have one half-set of chromosomes). Only females have fathers. It can be seen that sisters have a 75 percent shared genetic relationship, whereas mothers and daughters have only a 50 percent shared genetic relationship. Kin selection therefore favours the raising of fertile sisters rather than fertile daughters. Males have no such special relationships, and therefore do not form sterile worker castes. (Maynard Smith, J. *Scientific American*, 1978.)

and behaviour involving aid to others but which serve an individual's own reproductive ends.

Following Hamilton, other models were proposed for biological altruism: notably one called "reciprocal altruism," which takes place between nonrelatives and which depends on a kind of mutual back-scratching (Trivers 1971). More generally, using ideas and concepts of game theory, workers in the field were able to set these and other models in an overall Darwinian context, showing when certain reproductive "strategies" were likely to succeed and to lead to the evolution of physical and behavioural traits which enable organisms to survive and reproduce among their fellows, and when they were not. Particularly significant was the idea of a "reproductively stable strategy": a course taken by an organism because none other will benefit the organism more in the social situation within which it finds itself (Maynard Smith 1982). Biological altruism was now seen as part of a more comprehensive Darwinian picture.

Together with the theoretical ideas, the workers of the 1960s and early 1970s turned increasingly to detailed and long-term empirical studies, both in the wild and in experimental situations, showing how the new models function, where adjustments are needed, and how new directions are to be sought. (See Ruse 1999.) For all that he himself was marching to the beat of a (somewhat) different drummer than most of his fellows, a major figure was Edward O. Wilson, even then with fair claim to be the world's leading expert on the social insects. A man for whom interconnections and synthesis are the very lifeblood of intellectual advance, he took readily to the task of creating a coordinated integrated subject or discipline. Giving this vibrant new field – the study of social behaviour from a Darwinian perspective – its official name, Wilson authored the magisterial *Sociobiology: The New Synthesis* (1975). Going right through the animal kingdom, this work surveyed the theoretical models and ideas and showed how they were finding confirmation in the real world.

It is fair to say that, in the two decades subsequent to *Sociobiology: The New Synthesis* – and to *The Selfish Gene* (1976), a sparkling popularization by Richard Dawkins – sociobiology has come into its own as a full member of the Darwinian areas of scientific inquiry. New models, new hypotheses, new techniques, new findings, new studies, all have helped sociobiology to take its place alongside such fields as palaeontology, biogeography, and systematics.

Humankind

What has made sociobiology controversial has been its extension to our own species, to *Homo sapiens.* In this century, the study of humankind from a biological perspective has been muted and often under a cloud for several reasons: the territorial ambitions of the social scientists, for one, and the dreadful distortions of human genetics by the Nazis for another (Degler 1991). But nothing has deterred the sociobiologists, who have rushed in to claim that kin selection, reciprocal altruism, and related models are the keys to understanding human behaviour, particularly as it occurs in group or social situations. Marriage relationships, family structures, parent-children interactions, social customs, religious beliefs, power structures, and more have been subjected to sociobiological analysis (Betzig et al. 1987).

Controversial though it may be, let there be no mistake that human sociobiology – something today often hidden under innocuous-sounding names like "evolutionary psychology" – is part of the general Darwinian picture: selection working on features powered by the genes. By illustration, take the oft-praised work on homicide by the Canadian researchers Martin Daly and Margo Wilson (1988). They argued that homicide should follow sociobiological patterns, and were gratified to find that this generally seems to be the case. One would expect those humans with most to gain and least to lose from violence would be those most likely to commit homicide, and Daly and Wilson argued that the people who best fit this pattern would be young males with little or no stake in society. Figures collected by police and other agencies confirm this hypothesis: a finding made the more striking by the fact that the comparative figures hold across societies (e.g., American cities as opposed to Canadian cities) with very different murder rates.

Particularly significant is the fact that the greatest apparent anomaly turns out to be the most triumphant confirmation. Homicide figures within families show that there is a persistent and steady number of murders by fathers of their children: rarely by mothers, except the infanticide of the newborn, which is a special case calling for its own explanation. Surely this male violence is a direct violation of Darwinian principles, for parents should not eliminate precisely those who carry on their genetic heritage. Daly and Wilson hypothesized that perhaps the homicides are by stepfathers killing stepchildren. This would make sociobiological sense (the mothers are thereby freed to attend to the needs of the stepfathers's own biological children) and in fact is a common finding in the animal world (Hausfater and Hrdy 1984). Male lions and lemmings, to take two examples, kill off all of the young when they take over a female. This hypothesis about the involvement of stepfathers proved to be precisely true. A man is *one hundred times* more likely to kill a stepchild than he is to kill a biological child! Showing how this was no explanation after the fact, it was not until Daly and Wilson started asking their questions that police forces began distinguishing in cases of homicide between biological and social parents. Previously, the distinction had not been thought relevant (Figure 11).

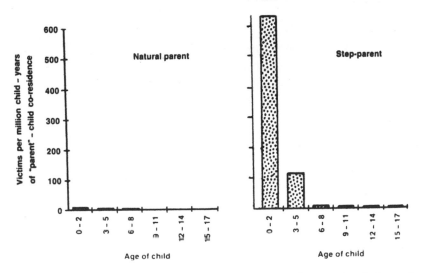

11. The risk of being killed by a stepparent versus a natural parent in relation to the child's age, (Daly, M., and M. Wilson. *Homicide*, 1988.)

The Evolution of Morality

For our purposes, without prejudice, let us now assume not only that sociobiology is a viable and trustworthy branch of science, but that this applies also to human sociobiology, however named. Pertinent to our inquiry is the fact that sociobiologists – myself prominently included! – have argued that we are in a position to make plausible suggestions about the evolution of human morality (Ruse 1986b). We start with the idea of altruism and with the division one must make between two senses in which this notion is used.

The key component in ethics – especially ethics at the substantive level – is the fact that we feel the obligation to promote (through our actions) the good, meaning that we feel the need to act kindly toward others simply because this is the right thing to do. This is altruism in the literal sense, and no one would deny, especially not the sociobiologist, that we have these moral sentiments and that they are genuine. Mother Teresa truly wanted to succour the poor of India because it was the right thing to do. The rest of us may not always be so good, but each and every one of us knows the tug of moral obligation. The other sense of altruism is what has been introduced as the biologists' sense, referring to the actions performed by an organism towards others because there is expectation (not necessarily conscious expectation) of return from an evolutionary perspective. This kind of altruism is metaphorical. It is not Mother Teresa–type altruism. It is ant altruism, and it is this which has been the focus of sociobiological attention. I will refer to it, in quotation marks, as "altruism." One should not confuse altruism (good done for its own sake) with "altruism" (help given for biological returns), but neither should one belittle either notion. A metaphorical understanding is not an incorrect or inadequate understanding. It is just not a literal understanding.

Now it is an empirical fact that humans have evolved in such a way as to be highly "altruistic," and moreover to be greatly dependent on such "altruism." These are not disconnected points, for there has obviously been evolutionary feedback. Humans are (compared to other mammals) not particularly strong or agile or fast or many other physical things. We need to cooperate to survive. Our Pleistocene ancestors could do little by way of hunting alone, unlike the lion or the cheetah. On the other hand, we are good at cooperating, and we have built-in biological devices against spoiling things through intragroup violence and strife. We do not have imposing weapons of destruction, like fangs or claws. And our hormonal balance keeps us all relatively calm. Hard though it may be to imagine, the murder rate among humans – even taking into account the mass killings of the last century – is less than that among many mammals.

The point is that humans are social and that we need social adaptations, like language and the ability to resist disease. Remember the sad tales of the isolated indigenous peoples who were killed by the viruses of those more used to living in large groups. Most particularly, we need to be "altruists," and we obviously are. But now the question comes of proximate causes. How in particular do humans put their "altruism" into effect? There are at least three possibilities, and in some respects humans have gone down all three routes.

Hardwired "Altruism"

The first possibility is that we might be hardwired, as one might say, by the genes to act in cooperative ways. We do what we do without any choice because this is the way of our biology. This is the cause of the "altruism" among the social insects. They are not thinking beings. They do not heed the call of the categorical imperative. They simply do what they do as automata, because their biology tells them so. And in some respects this accounts for human "altruism." Parent-child relationships are frequently of this kind. One responds instinctively to the needs of one's own offspring: especially in the case of mothers for whom (like other animals) a kind of imprinting takes place, bonding them to their children. Out of instinct, one loves one's own children more than the children of others.

Hardwiring is important, but there are very good reasons why such hardwiring – innately motivated feelings and actions, if you like – cannot account for the whole of human "altruism." Hardwiring is fine if nothing goes wrong, but if something does go wrong – the unexpected occurs – then you are in trouble. Take the ants. When foraging, they follow pheromone (chemical) trails. If it rains, then they can be lost forever. But it hardly matters to the queen (or to her sterile daughters, for that matter), for she produces literally millions of offspring. There are lots more to fill the gap. Humans have gone the other route of producing just a few offspring; but the cost is that we cannot afford to lose them carelessly. It would be a disaster if it were fatal every time a child got caught in a rain shower. There have to be more sophisticated mechanisms motivating and controlling humans, so that our actions, including our "altruistic" actions, can respond to change and challenges.

Super-brain "Altruism"

This suggests a second proximate mechanism, the very opposite of hardwiring. Perhaps to produce "altruism" humans have evolved as superbrains, calculating carefully the costs of any social interaction and acting positively only if it is in our self-interest. Selfish in the sense of looking to and only to our own needs and desires, although not necessarily selfish in the sense of grabbing more than our share. It is often thought that superbrains would be forces for evil – Darth Vader types – but this is not necessarily so, especially if everyone else is a super-brain. I want to take everything for myself, but I know that you want to take everything for yourself, so we have to compromise.

Again, humans have clearly gone down this route to some extent. When you purchase something from a store, you and the storekeeper are interacting socially: being "altruists" in the sense of doing things for the other for your own ends. You do not love L. L. Bean, nor does he love you, but you get on perfectly well together doing things for him (giving him money) as he does things for you (giving you goods). And if you do not like his goods, you can take your custom elsewhere, and if he does not want your money, he can stop offering the goods you want. But obviously we are not complete super-brains; this is not true even of professional philosophers. Perhaps part of the reason for this is physiological or a lack of evolutionary time. Super-brains might not be that easy to produce biologically. But there are other, good, selective reasons. Even super-brains are going to need time to calculate their self-interest. However, in life, time is money. You often do not have the luxury of infinite time to make decisions. A tiger approaches. You and your fellow human are in danger. Should you warn him? What will be the benefits? Will he give you something? Should you demand first? Are his promises reliable? By the time you have made your decision, both of you are in the tiger's stomach!

Super-brains are a bit like those early chess-playing machines. They thought of all the options, but were useless precisely because they thought of all the options! After a move or two, they were paralysed because there was so much to consider. What was needed was an approach which incorporated quick-and-dirty solutions: strategies which would generally win (based on past experience) but which could certainly be beaten, since they were not perfect. And now, of course, with Deep Blue, the machines have been improved to such a point that in real life even the very best chess players can be beaten. Which analogy takes us to our third option for achieving biological "altruism."

From Altruism to "Altruism"

We want a pragmatic solution, one that will generate the right moves most of the time, even though there will be mistakes and breakdowns and some actions will misfire. In the case of the machines, we have (as it were) certain strategies hardwired in, so that the moves will be made in a certain way given the initial conditions. The sociobiological claim is that in the case of humans we are genetically predetermined to think in certain ways, so that in specified situations we will incline to act in certain ways. And the genetic predetermination manifests itself as a moral sense: an awareness of certain rules or guides which are binding upon us – the prescriptions of normative ethics.

In other words, the claim is that in order to make us good biological "altruists," natural selection has made us moral altruists. And note that the claim is that this is a genuine morality. By nature, we are going to be selfish or at least self-serving. If selection did not make us this way, we would die out immediately. The person who has no concern for food and drink, no interest in sex (or who willingly steps aside in favour of a rival), may be a saint, but he or she is going to be a Darwinian flop. Yet, because we are social animals, we need something to break through this barrier, to make us interact with our fellow humans, to make us biological "altruists." And this something is going to be the moral urge: the feeling that we "ought" to do certain things, even though our nature is against it. Not that any sociobiologist wants to claim that biology does it all. It is always a function of biology in the environment: in the case of humans, of biology in culture. So one expects to find that there will be cultural differences across societies in time and place. If nothing else, different technologies and different challenges call for different solutions. But underlying it all is a shared moral base: the morality needed to make social animals biological "altruists."

Biological Normative Ethics

What of Christianity in the light of all of this? We are assuming that what has just been presented is well taken. It is a Darwinian empirical argument and in this context is not to be questioned. Our first question or set of questions must be about normative ethics. What has the sociobiologist to say about the details, and how do these compare to Christianity? In a reasonably straightforward sense, the mesh is going to be good. The sociobiologist is committed absolutely and completely to the genuine nature of human altruism. The causal process might make use of "selfish genes," to use Richard Dawkins's felicitous metaphor, but this does not imply selfish people. Indeed, the very crux of the sociobiological case is that we need real altruism to make us break through our usual selfishness. We need something as powerful as this, or "altruism" will not be achieved. For this reason, the sociobiologist endorses completely the Humean distinction between "is" and "ought" and thinks the naturalistic fallacy a genuine fallacy. Morality is different. "I love my children." "I ought to love my children." These are two quite different claims.

This confluence of Darwinism and Christianity continues as one gets into specific moral issues. Take sexuality. You are sadly mistaken if you assume that, because selection promotes adaptations for reproduction and breeding, Darwinism is going to be pro-sex in a way that Christianity is often taken to be anti-sex. I doubt that one can readily mount a Darwinian argument for celibacy, but certainly one can mount a Darwinian argument for restraint. I want to sleep with every pretty woman that I meet. That is my biological nature. But if I did, or tried to, chaos would soon obtain as every other male tried to do the same. Society would collapse. Like a stag with a harem, I would be so busy guarding my mates, making sure that no one else encroached on my sexual property, that I would do nothing else: until, like the stag, I dropped from exhaustion and starvation. But I do control my sexual impulses more or less. Partly out of prudence. If I started to crawl over the president's wife at a formal university dinner, I would soon be out of a job. But partly, and chiefly, I do it because I have the sense that it is right to exercise personal sexual restraint. I do not try to sleep with every pretty woman I encounter, because that would be wrong. This is not to say that I never do wrong, that I have never slept with someone else's wife. Sociobiology (and Christianity, for that matter) is not denying imperfection; it is rather trying to explain why we are not totally imperfect.

The Darwinian approach will not parallel everything that every Christian wants to claim in the name of morality. There is no ready sociobiological argument for or against abortion. No Darwinian wants every fetus aborted, but one can think of scenarios where abortion might make Darwinian sense: for instance, if a mother were to die and leave many young unattended children. Some Christians would find this conclusion unacceptable, although of course other Christians would probably argue in a way very close to that of the Darwinian. In the light of these differences, not so much between Darwinian and Christian but more between Christian and Christian, it is worth noticing something about the general nature of moral arguments. Rarely if ever is a conclusion drawn or decision made purely on ethical grounds. Almost inevitably it is ethics in conjunction with something else, usually beliefs or claims about the empirical world. Should I plunge a knife into this person's chest? Yes, if I have reason to believe that I can then clean out or remove blocked arteries. But not otherwise, and likewise with abortion. Much of the debate is about the status of the fetus, the mother's happiness now and in the future, and so forth. Both Darwinian and Christian think that needless killing is a moral wrong. The weasel word here is "needless," and much of the abortion debate is about matters of empirical fact around our understanding of this term.

What of the Catholic concern with natural law? The Darwinian no less than the Christian will be concerned about judging what is natural, and sympathetic to the claim that what is natural is what we value. The Darwinian is a naturalist, as was that good biologist Aristotle, to whom so much Catholic thought is indebted. This does not preclude genuine differences about what is "natural." Remember the issue of birth control, with the Darwinian suggesting that intercourse maintains pair bonding and the conservative Catholic arguing that its only function is reproduction. Nor does it deny that, at a deeper level, disagreement may come over the precise connection between the natural and the morally correct. The Darwinian may find the biologically unnatural aesthetically displeasing, but moral judgements will probably require more. Suppose one thinks homosexual activity unnatural. The natural-law theorist will at once judge it immoral – it is a violation of God's eternal law. But even though the Darwinian may find it unsettling at a personal level, this is not to say that it is wrong. In order to make the moral judgement, one needs something more (Ruse 1988b). And this something more will involve reference to the basic moral standards to which one subscribes. One cannot simply equate the natural with the good. (If, as a Christian, Catholic or Protestant, one insists that the natural must be equated with the good, then in order to bring one's thinking in line with biology, one can argue that even though one might find something unsettling, this does not in itself affirm that it is unnatural. Providing that extra element, showing that the personally unsettling may not be the objectively unnatural, is the motive for those appeals to homosexual activity in other cultures and among other animals.)

Supreme Principles

Turn to the basic or supreme moral principles: the supreme principles of substantive ethics, that is. What kind of "altruism"-promoting altruism is one going to get from natural selection? Hardly any surprises, I suspect, although it is probably wrong to seek one and only one principle of morality binding on all people. Because one is taking a naturalistic approach, one expects a range of emotions and obligations, within certain limits. One is going to expect a kind of commonsense morality, with an underlying base of reciprocation: reciprocation because it is right, not because I have done something for you. "Be kind to people." "Help children and the less fortunate, and try to do so in proportion to need." "Give priority to mothers." "Don't rape and/or use gratuitous violence towards women." "Keep your word." "Do not take what is not yours." "Try to moderate habits like boastfulness which are going to irritate others." "Stick up for your country or your group." And so forth (Mackie 1977, 1978).

Philosophers like to isolate one or a few unique all-encompassing moral principles – like the categorical imperative – and then defend their choices against those of others, devising strange examples which favour one side or the other. There is no real harm in doing this sort of thing. After all, it is our job. But it should not blind us to the fact that although most people are not moral philosophers with a clearly articulated system, they do fairly well nevertheless, and that all moralities which have stood the test of time (thus excluding perversions like Nazi morality) concur on the basics. If you insist on having a system, then the social-contract scheme of John Rawls (1971), the most famous American moral philosopher since the Second World War, seems to fit fairly well into the sociobiological picture (as Rawls himself admits).

Rawls would have us be just. This does not mean giving everyone exactly the same. It means rather giving according to people's needs and wants in such a way as to maximise the benefits for all. Inviting us to put ourselves behind a hypothetical veil of ignorance about the position we will have in society – born of rich or born of poor, born talented or born stupid, born healthy or born sick – Rawls suggests that the just society is the fair society. This is a society where your gains are the best possible consistent with everyone else getting their best possible. If you are talented, then you might want to get more than the stupid; but, behind the veil, you have no way of knowing where you will be on the talented/stupid scale. Hence, it is hardly in your interests to give more to the talented just because they were born lucky. You might be born untalented. But good doctors, for example, are beneficial to everyone, so if the only way to get the talented to take up the onerous burden of medicine is by paying them more than we pay philosophers or garbagemen, then we all benefit, talented or stupid. Such inequality is consistent with justice.

This is a kind of social contract where society is seen as an agreement to cooperate in order to obtain the benefits which we get out of cooperation. The nice thing is that, if you introduce sociobiology into the picture, instead of pretending that a group of ancestral elders got together to set up the rules – which did not happen anyway and would not account for the sense of obligation we have about morality – the burden of starting everything off is put on the very real genes as worked on by very real natural selection (Ruse 1986a). I hardly need say how all of this – whether you remain with commonsense maxims or opt for a system like that of Rawls – fits well with Christian prescriptive morality. The morality of the Ten Commandments is right in line with commonsense morality, and the love commandment lays the way for direct reciprocation. Love your neighbour as yourself, but although you must forgive your neighbour for not loving you, you have the right – many would say you have the Christian duty – to see that your neighbour reciprocates, to you and to others. Not because you and they have given, but because it is "right."

Of course, historically, you would expect the Christian to feel comfortable with Rawls, for he acknowledges explicitly his debt to Kant, and as we know Kant in turn was deeply influenced by the Pietism of his parents. It will not have escaped the reader how very Kantian-sounding was my discussion of why the sociobiologist would never think that unbridled sexual promiscuity is supported by Darwinism. If everyone attempted it, then we would soon have a full-blown societal Kantian contradiction.

Morality's Range

Sociobiological ethics meshes with Christian ethics. But is this not all a little bit too smooth and optimistic? Does not Christianity try to push you out and beyond the commonsense maxims embraced by Darwinism? Jesus was not addressing your average well-fed member of the Rotary, who does his bit for handicapped children and who then goes home well satisfied to enjoy the fruits of his business. Or rather, he *was* addressing such a person and saying that this is not enough. "Turn the other cheek." "Give all you have to the poor and follow me." "Think not for what you will put on . . ." and much more. A naturalistic account of morality like that of the sociobiologist may go so far, but ultimately it cannot go as far as the Christian demands in the name of the Lord.

This is a serious objection, and it should not be minimized. It is indeed true that the sociobiological substantive ethic is going to be limited. Reciprocation only works between those who can reciprocate. Self-benefit does not demand that every recipient give something immediately in return. Your contribution may be like an insurance policy payment, that is, something never returned because you never need to draw from the common pot. And it may be that you or a relative may never be able to contribute, because of illness or whatever. Hence, you are (if able) willing to give to those in your group who can never reciprocate. "There but for the grace of God go I." But self-benefit does mean that your social fellows have to be in the same pool as yourself. People beyond the pale, people with a different insurance company, cannot expect to draw from you, from your policy. In less metaphorical language: you expect to find morality falls away as one leaves first the family group, and then the immediate social group, and so on out, to the country as a whole. David Hume, an enthusiast for a naturalistic ethics, spotted this point 250 years ago. "A man naturally loves his children better than his nephews, his nephews better than his cousins, his cousins better than strangers, where every thing else is equal. Hence arise our common measures of duty, in preferring one to the other. Our sense of duty always follows the common and natural course of our passions" (Hume 1978, 483–4).

In this day and age particularly, this does not mean that you can and should be indifferent to the starving poor of Africa. Apart from anything else, there may be expediency reasons for worrying about them. Social diseases, for instance, have a nasty way of becoming worldwide. More pertinently for morality, in the age of television and jet travel and e-mail, one's group stretches out to the whole world, in some sense. But it does mean that you have a bigger obligation to your own children and those in your local group than you have to the children of Africa. It also means that your sense of obligation to animals will be truncated. They cannot reciprocate in the way of other humans, and so one rather downgrades the sense of obligation felt towards them. One does have an obligation, but a limited one. Cruelty is wrong. Meat eating is an option.

Many Christians would recoil at this point, finding these sentiments quite unacceptable. Loving your neighbour as yourself means loving everyone, everywhere, at every time (Singer 1981). No one pretends that we do always or perhaps ever do this, but the Christian is only too aware of sin. We ought to do this, and inasmuch as we do not, we fail God. Jesus himself is positively brutal on the question of the family. He brushes off his mother and his brothers, he is contemptuous of a man who wants time to bury his father, and he drags his disciples away from their families. "I have come to set a man against his father, and a daughter against her mother, and a daughter-in-law against her mother-in-law; and a man's foes will be those of his own household" (Matthew 10:35).

We may have an insoluble problem here, although it must be noted that we have a difference dividing Christians among themselves, rather than all Darwinians from all Christians. While some Christians insist that one should interpret the love commandment in the fully universal fashion, this is far from a general sentiment (Wallwork 1982). There is a tradition which states that charity begins at home and that we truly do have differential obligations – and that to pretend otherwise is itself immoral. How many of you would think me a saint if you learnt that while I give three-quarters of my income to Oxfam, my family lives in a two-room shack and rummages for clothes at the local Salvation Army thrift store? Remember *Bleak House*. Dickens rounds savagely on Mrs Jellyby, concerned as she is for the benighted heathen of Borrioboola-Gha, because she neglects her own family and the desperate plight of Jo the crossing-sweeper.

There is biblical warrant for those who argue for a restrained interpretation of the love commandment. One must remember always that Jesus thinks and speaks and acts in an apocalyptic fashion, concerned for the immediate present rather than for the long haul. Notwithstanding other comments, Jesus is supportive of monogamous marriage, setting strict limits on when it might be dissolved; he shows concern for his mother when he is on the Cross; and his followers certainly imply that one has family obligations which should be taken seriously. "If anyone does not provide for his relatives, and especially for his own family, he has disowned the faith and is worse than an unbeliever" (1 Timothy 5:8; this is written in the name of Paul but probably composed by another writer after his death).

What is really important is that Christ was a preacher, not a philosopher. Whereas sociobiology tells us how and why we feel about morality, Christianity is trying to pull us out of our moral complacency toward better things. You are never going to effect much change if you simply compliment people on what they are doing. Jesus was dragging us to and beyond the limit. The sociobiologist is trying to do more than merely describe feeling. I did not need sociobiology to tell me that I love my children more than I love the children of others. Sociobiology, rather, is trying to explain moral obligation: I feel an obligation to my children that

I do not have towards others. The Christian, through Jesus, is trying to enforce and extend these obligations. We forget that the obligations do extend beyond self and family. We slide readily into selfishness. Christ's message is intended to lift us up and beyond ourselves. It is easy to do right to our own children. It is a lot less easy to do right to the children of others. This is why Christ spoke to us as he did.

Metaethics

What of foundations? As with social Darwinism, the question of biological progress is important. Refer now, as before, to natural forces based on selection, without implying the need for either non-natural elements or significantly non-Darwinian mechanisms. If you think that humankind came about through progressive forces for change, then since (substantive) morality is so essential a part of our nature, for the Christian there is no real difficulty at the metaethical level. One is not committing the fallacy of getting "ought" from "is." One simply believes that morality is underpinned by God's will, and that evolution is the way in which it came into effect. The fact that the moral sense is an adaptation is neither here nor there. Eyes are adaptations, but the train I see bearing down on me is real; likewise the moral sense is an adaptation, but this does not preclude its appreciation of the genuine nature of God's will.

In fact, as with social Darwinism, evolutionism rather clarifies and justifies the Christian position than detracts from it. God wants us to love each other, but none of this is simply a question of God's unbridled whim. He is concerned that we express our humanity – which reverts ultimately to having been made in His image – in the fullest and most fruitful possible way. Since God has made us as social animals, this means above all else being in loving friendship relationships (agape) with others: it also means, if we are fortunate, being in loving sexual relation (eros) with another (qualifying clauses about the virtues of celibacy for the priesthood and so forth being taken into account). Love is not just an arbitrary notion of God. Although we can all be hateful, for God to insist that we should all hate each other (and think we should hate each other) is just not on. Even (especially) the most intimate of sexual relationships and actions, giving completely to another and being so totally exposed (emotionally and intellectually, as well as physically), contribute to our wholeness and health as functioning moral and social beings. People who have never risked making fools of themselves in love are people who have a truncated relationship with all fellow humans.

We seem to have circled back to a natural-law type position, and this is surely true. What God has produced through evolution is good – better than what was before – and it is our obligation, either on a case-by-case basis or through our general principles, to cherish and enjoy and respect it. Sexuality in itself is a good thing, and inasmuch as we use it properly we are "participating" in the eternal law. The Darwinian who is a Christian justifies his or her position here by reference to the way in which God has made things of positive value through the natural, progressive system. Doing things which are natural is not right simply because these things are natural, but because the natural is good as intended by God. This does not mean that we are frozen into a conservative morality, as many Thomists think. If you can show that modern biology demands rethinking empirical claims (say, about birth control), then one's moral position might likewise be rethought. Or one might show that humans, thanks to their biology (big brains and so forth), escape the strict confines of biology and move into culture. Hence, judgements of naturalness should no longer be based purely on morphology. (Is it immoral to get into an aeroplane and fly, despite the fact that we have no wings ourselves?) But it does mean that one's morality is being constrained and defined by what one takes to be the process and product of evolution.

What if you deny that evolution is progressive? What if (since you are now *ex hypothesi* a Christian and think that humans are special) you have interpreted your science in an Augustinian fashion, arguing that God's intentions come through (Darwinian) processes, even though there was no progressionism as such. One thing you have to accept is that evolution could have produced beings very different from us – and, if you accept the plurality of worlds, possibly has produced beings very different from us. You might have rational beings who did not at all have the same social structures as we, and who did not have the same moral sense either. Suppose women (or "women") came into heat in some coordinated way, and there was something like Oktoberfest in Munich, with unrestrained copulation for a limited time. Such beings would certainly not have the same sexual mores as our own. Would promiscuity be immoral under such a situation? I am not sure that the answer is yes.

Even the deep fabric of morality might be changed. Such beings might be social and yet not recognize the love commandment or anything comparable. Certain formal structures might have to be obeyed: I give you something, you give me something in return; I give you something, you fail to give in return, I give you one and only one chance to redeem yourself – that sort of thing (Skyrms 1996). And how we are motivated might be quite different. Think of what I will call the "John Foster Dulles system of morality," named after Eisenhower's secretary of state at the height of the cold war. He hated the Russians; he thought it was right to hate the Russians; but he knew also that they hated him – so there was balance and compromise. Why should we not have a general system of morality which works this way, with cooperation fuelled on fear and hatred – fear and hatred of a moral kind – rather than love? What price God's will now?

For the Christian moralist, relativism is anathema. One can certainly accept that different societies may well have different customs, but there has to be an underlying universality to morality. We are all made in God's image, and there cannot be one rule for one set of people and another rule for another set. There are "judgements of moral conscience, which Sacred Scripture considers capable of being objectively true" (John Paul II 1998, 82). Suttee (the widow joining her husband on the funeral pyre) is wrong, whether it be in Britain or in India, in this century or the last. The sociobiological account of morality is in agreement about relativity here on Earth with respect to our species. Morality has to be something shared or it will not function, and inasmuch as it is biologically based, since we are all the same species there probably is not much variation. But we do now seem to be faced with an intergalactic relativism.

Probably the Christian will think that if this is the greatest threat that Darwinism can pose to Christian ethics, there is not much need for worry. Let us wait until we meet rational aliens. Or the Christian might point out that, even under Dulles morality, there may still be a place for shared virtues: sticking to one's convictions, for instance, whatever these convictions may be. In this respect, God lays the same mandates on us all. All I say is that if one rejects progressivism, then one has an added task in trying to harmonize Darwinism and Christianity. It is not necessarily impossible, but it is a task which will need to be performed. The Darwinian can be a Christian, but both sides have to think about their absolutely bottom-line commitments, and about where and how they might be prepared to compromise or show flexibility to achieve harmony.

CHAPTER TWELVE

Freedom and Determinism

I come to my final topic. Let me recap the Christian position and then see what Darwinism has to say on the subject and how the two compare.

Original Sin

Freely, God created Heaven and Earth and put us humans in a privileged place within this creation. It is an absolutely central part of Christian theology that we humans likewise are free agents. We ourselves have the power to evaluate and decide between courses of action, and to act on our own decisions. This is at the very heart of what it means to be made in the image of God. But since we have free will and are created by God, how then is it that we sin? How is it that human-caused evil comes into existence? We could not sin unless we were free – the earthquake is not sinful, no matter how many it kills – yet why do we misuse our power, if a perfect Being created us? Surely He cannot be the source of sin?

No indeed! Sin comes from within, and since there is no apparent reason why we should be innately sinful, it is here that the notion of *original sin* comes into play. The first humans – Adam and Eve – freely chose to disobey God's explicit command. Even those who accept today's science insist on some point of moral failure: "In the course of evolution, there must have been a first moment of conscious moral choice. That is the point at which the 'fall of humanity' began and humans were estranged from that natural fellowship with God which should have been theirs, and from their natural ability to relate unselfishly to one another" (Ward 1998, 42). From here on, down through all humans, the corruption persists. It is not that we are born actually having sinned personally, but through the act of sexual intercourse (to take the traditional position articulated by Augustine), the taint is passed on. It is like a hereditary disease, a power which holds us captive, or a form of guilt which we all have. Ultimately, we are saved only by God's grace. For Catholics, the stain of original sin is washed away by baptism, but for both Catholics and Protestants an inclination to sin, a "concupiscence," remains.

This means that, as we choose, we are burdened or weighted or biased towards doing the wrong thing. We have free will, so we can choose the right thing, but original sin makes us liable to go wrong. Augustine's image is of a balance. If the balance be true, then we choose right or wrong without distortion. "But what, asks Augustine, if the balance pans are loaded? What happens if someone puts several heavy weights in the balance pan on the side of evil? The scales will still work, but they are seriously biased toward making an evil decision." This is exactly the case with human nature and original sin. "The human free will is biased toward evil. It really exists and really can make decisions – just as the loaded scales still work. But instead of giving a balanced judgment, a serious bias exists toward evil" (McGrath 1997, 427–8).

The Pelagian heresy sets off and highlights the Christian position. For Pelagius, we are totally free and able to do the good. To imply otherwise is to compromise God's integrity: implying that He did not create things perfectly or that He intervenes in our choices. It is for this reason that the Pelagian has us judged by God according to our actions, whereas the Christian (Augustine, strongly followed by Luther) has us saved by God's grace, irrespective of our actions, which in any case can never balance our sin. This does not mean that good works are irrelevant to the Christian, but rather that they are marks of our worship of God and acknowledgement of His goodness and power. They are consequences rather than entry tickets. "For as the body apart from the spirit is dead, so faith apart from works is dead" (James 2:26).

Free Will and Predestination

The notion of freedom raises one of the biggest problems in Christian theology. It is a central belief of Christianity – one developed by Augustine and taken up by the reformers – that God, being omniscient, can foresee the future, including our future actions and His responses to them. In this sense, our lives are "predestined": with the consequence (particularly stressed by Calvin) that some are born sheep and destined for salvation, while others are born goats and destined for hell. Obviously, none of this bodes well for free choice. A predestined sheep will be sheeplike and a predestined goat will be goatlike. What price free will now?

There is a standard response. The opposition for the Christian – for Augustine (1972) particularly, who, having raised the issue, found the solution – is not between freedom and inevitability, but rather between autonomy and fate. We humans are free in the sense that God does not rule our lives, making certain events unavoidable: as Oedipus was destined to kill his father and marry his mother, no matter what he or anybody else did. We are autonomous beings who can make our own choices which do make a difference. (For this reason, the church has opposed astrology, which does imply that our lives are ruled by fate.) We are therefore open to God's gift of grace and liable for punishment. But this is not to deny that our actions can be understood and foreseen: foreseen in their entirety by God, who therefore knows what will obtain.

In explaining and resolving this paradox of predestination, Augustine uses God Himself as an analogy. He is free, supremely so, and yet when faced with a choice we know what He will do. He will do the good. By His nature, He can do no other. Likewise, God knows what we will do when we are faced with choices. We have free will, but God knows what we poor sinful beings will do. The logic is the same as that used against the existence of natural evil. God creates freely and does so in perfect love. But He cannot do the logically impossible. Likewise He chooses freely, but it is logically impossible that He, a perfect being, will choose the wrong.

We do not put the life of God or the foreknowledge of God under necessity if we should say that it is necessary that God should live for ever, and foreknow all things; as neither is His power diminished when we say that He cannot die or fall into error, – for this is in such a way impossible to Him, that if it were possible for Him, He would be of less power. (Augustine 1950, Book V, chapter 10, 157)

The same is true of us. We choose freely, and God can know what is going to happen without at all compromising this choice. "When we say that it is necessary that, when we will, we will by free choice, in so saying we both affirm what is true beyond doubt, and do not still subject our wills thereby to a necessity which destroys liberty" (Augustine 1950, Book V, chapter 10, 157). We, imperfect as we are, will choose as we choose according to the balance between our God-given nature and our taint through original sin. We are autonomous, but God can forecast what we will choose. "It is not the case, therefore, that because God foreknew what would be in the power of our wills, there is for that reason nothing in our wills. For he who foreknew this did not foreknow nothing." In short, you can have things both ways: freedom and divine foreknowledge. "We are by no means compelled, either retaining the prescience of God, to take away the freedom of the will, to deny that He is prescient of future things, which is impious. But we embrace both. We faithfully and sincerely confess both." Hence, notwithstanding foreknowledge, God can and does rightly punish us for our transgressions – except where freely He chooses to save us through His grace. We have no right to that grace and get it only by acknowledging His power and our fallen state and begging Him for His gift.

Augustine is thinking in a theological context. The scientific revolution raises some of these issues again, in a scientific context. Perhaps fate continues to rear its ugly head, threatening freedom: only now we speak of the restraint as "determinism." Could it not be that the laws of nature apply throughout, to the whole of creation including humankind? In which case, are we not mere automata, without any genuine freedom? As the ball falls inevitably according to laws, so also the human acts inevitably according to laws. This means, apparently, that there can be no such thing as free will, for that notion seems to imply a choice by humans: that we could, if we wished, adopt alternative courses of action, and hence that the future could be other than it actually is or becomes.

In fact, many philosophers (following Hume) are "compatibilists," arguing that far from determinism preventing or excluding free will, the two notions are compatible. One can have free will and determinism (Ayer 1954). Indeed, most of these philosophers would argue that unless one has determinism one cannot have free will! First, one distinguishes between autonomy and constraint. The meaningful sense of free will is not denial of determinism, but autonomy: being able to do something without external constraint. The prisoner in chains is constrained and thus not autonomous and hence has no free will, whatever the laws of nature may be playing at. Second, the converse of determinism is indeterminism, things happening randomly by chance. But no one wants to argue that the person who acts randomly or by chance – erratically – is truly free, let alone uniquely free. Such a person is mad, not responsible, which is the burden of freedom.

Augustine is controversial and Hume even more so. But, for the moment, let us see what Darwinism has to say on the subject. We shall return to the philosophy.

The Biology of Original Sin

Start at the beginning with the doctrine of original sin. Problems which arose before, arise again. If one is an evolutionist of any kind, then the Adam and Eve story must be modified in some respect. Even if one is happy with the Augustinian story of inherited guilt from Adam's action – and there are many Christians who feel acutely uncomfortable with this story (why should I be blamed for the actions of someone else?) – the Garden of Eden scenario, with the lion lying down with the lamb, sticks in the craw of the evolutionist. And the whole business of an original, unique Adam and Eve goes flatly against modern evolutionary biology. It is the old problem of souls. Is one supposed to believe that the parents of Adam and Eve – for they will have had such in the evolutionary story, if not in Genesis – were soulless or sinless or what? And what about their brothers and sisters, and the next generation of *Homo sapiens*, most of whom were not descended from Adam and Eve? Did God let humans evolve and then intervene with just one couple: at the very least changing their nature or behaviour, making them open to moral judgement? On the evolutionary story, their predecessors were certainly committing things which look very much like sins.

There is no call for despair. If one is prepared to accept a metaphorical interpretation of the Adam and Eve story, while insisting on the truth and the relevance of evolutionism – of Darwinism, in particular – a ready understanding of original sin offers itself. All of the elements have been presented to us in the last two chapters. The Darwinian, the sociobiologist, starts with the struggle for existence and reproduction and the consequent selection of variations leading to adaptations designed for success in this struggle. Many of these, as Thomas Henry Huxley and others have always stressed, involve self-interest, if not outright selfishness, with the host of features and attitudes and characteristics that we all find offensive and that the Christian judges sinful. To be self-interested is not necessarily to be immoral. No one judges ill the person who eats a meal because he or she is hungry, or who falls in love with a pretty girl or handsome young man and wants to have that person as a mate. But, all too quickly, self-interest runs into qualities like greed and lust and boastfulness. There are good biological reasons for this. The man who feeds himself (or his family) is better off than the man who has no food or just some leftover scraps. The man who impregnates a thousand women is ahead (in the Darwinian game) of a man who impregnates just one. The man who cheats and lies his way to the top of the greasy pole is more successful than he who loses, and who dies in abject poverty.

Original sin is part of the biological package. It comes with being human. We inherit it from our parents and they from their parents: they acted as they did, and because they acted as they did, it is passed down to us. Moreover, if you accept that God could not have created without the laws being as they are, He is blameless for the way things are. It is not His direct fault that we are sinful or that this is a tendency which we inherited. But of course, the point for the sociobiologist – and here the Christian's ears should prick up – is that this is not all that there is to the story. Laid on top of our selfishness is our (genuine) altruism, put in place to make us efficient biological "altruists": a very necessary adaptation, given that we humans have so firmly gone the route of sociality. We are loving and kind and generous – really loving and kind and generous – because that is just as much a part of our nature as is our selfishness. Eating all of the supper or impregnating a thousand women may seem like a good idea, but apart from the fact that others are going to have a say in the matter, we ourselves have feelings – moral feelings – that such actions are not good or wise or right. We may not always obey the call of conscience, but it is there nevertheless.

With respect to original sin, sociobiological *Homo sapiens* is nigh identical to Christian *Homo sapiens*. They both see humans as deeply self-centred, selfish even, but with a genuine moral sense overlaid on this: guiding (at least, instructing) our actions in social situations and interactions. The surface stories are very different, but the underlying concerns are the same exactly. According to both pictures, humans are truly sinful, with goodness fighting for control. (For more discussion on how today's theologians are wrestling with original sin in the light of modern science, see Korsmeyer 1998 and Oakes 1998.)

Genetic Determinism?

Turn now to the question of free will. Predestination is a Christian problem, and the Darwinian can sit out the theological niceties of this issue. If
the Christian does not like Augustine's solution, then the Christian must
find another. Darwinism, of course, must not conflict with – perhaps will
even support – whatever solution is found acceptable. But since we are
here dealing with a problem more theological and philosophical than
scientific, let us shelve this for the moment. Alternative theological
solutions – for instance, the universalism first mooted by Origen in the
third century, and endorsed by Karl Barth in the twentieth century, that
all sinners will eventually be saved – do not pose greater difficulties for
Darwinism than the Augustinian position.

Determinism is a more directly science-inspired belief. It makes better sense to start discussion of Darwinian connections at this end of the
spectrum, especially since it would be craven to deny that Darwinian
sociobiology (if well taken) is a piece in the case for determinism – just as
human sociobiology is for human determinism (Ruse 1979b, 1988b). The
whole point of human sociobiology is to show that there are certain
causes – events from the past – which govern or control or influence our
behaviour today. Why do young men want to sleep fairly indiscriminately
with young women? Why do young women decline most such offers?
Because of things which happened before we were born. It is as simple as
that. At some level, it is the presupposition of sociobiology that a deterministic account of human nature can be given and that sociobiology
contributes to this. The genes do not do everything – culture is also
important – but the genes are key players. Remember: "What, we are
then compelled to ask, made the hypothalamus and limbic system? They
evolved by natural selection. That simple biological statement must be
pursued to explain ethics and ethical philosophers, if not epistemology
and epistemologists, at all depths" (Wilson 1975, 3).

Full-blooded genetic determinism precludes freedom and any possibility of morality. Dan Dennett gives a beautiful case of such genetic
determinism, which he calls (after its chief player) "sphexishness." A wasp
(*Sphex*) digs a hole, finds and brings in a cricket which she stings to
paralyze but not to kill, lays her eggs next to this food store, and finally
closes off the hole never to return. A wonderful case of thoughtfulness

and intention, until something goes wrong and the mechanical nature of the whole process is revealed. "The wasp's routine is to bring the paralyzed cricket to the burrow, leave it on the threshold, go inside to see that all is well, emerge, and then drag the cricket in. If the cricket is moved a few inches away while the wasp is inside making her preliminary inspection, the wasp, on emerging from the burrow, will bring the cricket back to the threshold, but not inside, and will then repeat the preparatory procedure of entering the burrow to see that everything is all right." This can go on and on indefinitely. "The wasp never thinks of pulling the cricket straight in. On one occasion this procedure was repeated forty times, always with the same result" (Dennett 1984, 11, quoting Wooldridge 1963, 82).

But already we know full well that the sociobiology of humans goes beyond the crudely deterministic. We are not marionettes dancing blindly to the tune of our DNA. That is the fate of the hymenoptera. The whole point about morality, and the reason why full-blooded genetic determinism will not work – why such genetic determinism on its own would not be adequate – is that we humans have taken an evolutionary route for which such simple determinism would be fatal: even the more complex genes-working-out-in-the-context-of-their-environment genetic determinism. In Dennett's language, we humans cannot afford to be sphexish. We raise too few children and put too much effort into them to risk losing them when something simple – like moving a cricket – goes awry. And this, as we know, is where morality comes in.

But does the sociobiological account of morality do enough for the Christian? Does it give a sufficiently enlarged dimension of freedom? Continue with the kind of analogy which would appeal to someone like Dennett. Ants or wasps – the hard-line genetically determined – are like cheap rockets. (Are any rockets cheap? You know what I mean.) They can be mass-produced and are effective so long as the goal is never altered and no disruption occurs. They have no mechanism for self-correction. If they miss their target, who cares? There are always lots more where they came from. Humans are like expensive rockets or torpedoes. They have goal-directed mechanisms built in. They can correct for changed targets or disruptive conditions. They are not guaranteed to succeed, but often they do. Which is just as well, because you cannot afford to produce very many.

These expensive rockets are causally determined. If you knew everything then you could predict their future direction. Yet clearly, they have a dimension of freedom, an autonomy, not possessed by the cheap rockets. Consider again Dennett:

Contrary to the familiar vision . . . , *determinism does not in itself "erode control."* The Viking spacecraft is as deterministic a device as any clock, but this does not prevent it from being able to control itself. Fancier deterministic devices can not only control themselves; they can evade the attempts of other self-controllers to control them. If we are also deterministic devices, we need not on that account fear that we cannot also be in control of ourselves and our destinies.

Moreover, *the past does not control us.* It no more controls us than the people at NASA can control the space ships that have wandered out of reach in space. It is not that there are no causal links between the Earth and those craft. There are; reflected sunlight from Earth still reaches them, for instance. But causal links are not enough for control. There must also be feedback to inform the controller. There are no feedback signals from the present to the past for the past to exploit. Moreover there is nothing in the past to foresee and plan for our particular acts, even if it is true that Mother Nature – gambling on our general needs and predicaments – did, in effect, design us to fend quite well for ourselves. Far from it being the case that we are completely under the control of our ancestors or our evolutionary past, it is rather the case that that heritage has tended to set us up as *self*-controllers – lucky us. (Dennett 1984, 72)

Levels of Desire

For the moment, let us keep pushing this line of argument, seeing if we can get a little more sophisticated. Even if we grant that a determined being can have a dimension of freedom, there is still the question of how exactly this freedom operates, given all of the various desires and ends that we have. One of the most powerful and cited of recent discussions of free will analyses the notion in terms of levels. At the lowest level, we have basic desires: for sex, for food, for a cigarette, to give something to a crying child. At the next level, we can also have desires, but more importantly we have wishes or "volitions." It is at this level that we get the notion of the "will."

It is my view that one essential difference between persons and other creatures is to be found in the structure of a person's will. Human beings are not alone in having desires and motives, or in making choices. They share these things with the members of other species, some of whom even appear to engage in deliberation and to make decisions based upon prior thought. It seems to be peculiarly characteristic of humans, however, that they are able to form what I shall call "second-order desires" or "desires of the second order." (Frankfurt 1971, 6)

What does this mean exactly? That we have desires at one level, and also, at a higher level, desires or wishes about these lower-level desires. "Many mammals appear to have the capacity for what I shall call 'first-order desires' or 'desires of the first order', which are simply desires to do or not to do one thing or another. No animal other than man, however, appears to have the capacity for reflective self-evaluation that is man-ifested in the formation of second-order desires." An organism that is free, therefore, is an organism which has second-order volitions (on this earth, humans uniquely), which it can satisfy according to its first-order desires. You want to quit smoking (second-order volition). You have a desire to smoke and a desire not to smoke (first-order desires). You quit smoking (you exercise free will).

A person's will is free only if he is free to have the will he wants. This means that, with regard to any of his first-order desires, he is free either to make that desire his will or to make some other first-order desire his will instead. Whatever his will, then, the will of the person whose will is free could have been otherwise; he could have done otherwise than to constitute his will as he did. (Frankfurt 1971, 18–19)

It is here that we have a place for moral responsibility. I have an urge to seduce the girl. I have a counter-urge not to tell yet another phony story about respecting her mind above all else. My second-order volition is that it would be wrong to sleep with the girl. If I myself can bring my urges in line with my volition, I am free, and not otherwise. "I was seized by sin and fell, despite my best intentions." "I really didn't want to betray my wife, but on the business trip I found myself spinning the same old line with this chick I met." Sometimes we have a real clash about the second-order principles themselves. I think it would be wrong to sleep with the girl, but I really do not care about anything but my own personal sexual satisfactions. Who bothers with dated moralities? Here we might have a third-order principle – maximizing happiness, for instance – or perhaps we are too shallow to go further and care not for how second-order conflicts are resolved. Such a person is not so much morally wrong as morally irresponsible: what Frankfurt calls a "wanton."

Now all of this is tailor-made for the sociobiological/Dennett kind of analysis. As the chess-machine example of the last chapter made clear, one has second-order principles or strategies guiding what is happening at the first order: I can move this piece (first order); I can move that piece (first order); in order to win one ought to move the first piece (second order). Likewise for morality: I can do this (first order); I can do that (first order); in order to do the right thing one ought to do this rather than that (second order). Of course, the second-order principles or strategies are hardwired – genetically determined, if you will – but this is no bar to freedom. Only Sartre at his most barmy ever said that we have a choice about the content of basic morality. The choice is whether or not to sleep with someone who is not your wife; not whether sleeping with someone who is not your wife is moral or immoral. Being moral and free means choosing to follow or not to follow morality, not choosing the morality itself. This is true notwithstanding the fact that if one has second-order conflicts, it may well be that one needs some moral training in order clearly to discern the nature of morality. This is part of the genes plus the environment making the human. No one ever says that we are going to change the morality itself, although we may well want to change people's thinking about morality. As in: "Feminists and blacks are trying to make older white males see how their opposition to affirmative action is morally misguided."

Christian Freedom

Dennett somewhat smugly labels his analysis as having provided a notion of free will "worth wanting." But is it enough for the Christian? There will be those who will demand a freedom outside of or beyond law, whatever the counterarguments (Van Inwagen 1983). Unless one is prepared to suspend the Darwinian framework, one simply cannot satisfy these people. There will be those for whom any line of thought articulated by Hume and burnished by Dennett will be anathema. "Enemies' gifts are no gifts and do no good." One simply cannot satisfy these people either. But for those able to disregard Sophocles' warning, who (like myself) welcome a form of compatibilism, we surely have a dimension of freedom which can be used by the Christian. Even the Augustinian can be satisfied. Humans are rational; they are guided by desires; they do have the call of volition; these volitions most centrally include morality; we are free or autonomous (ceteris paribus) to satisfy this morality, either through these desires or by putting these desires into action.

At the same time, we can see the operation of original sin. Sometimes we are too much in the thrall of our first-order desires to obey or satisfy our second-order moral volitions: I wanted to stay sexually pure but lust overwhelmed me. As Saint Paul put it, "I do not understand my own actions. For I do not do what I want, but I do the very thing I hate" (Romans 7:15). Sometimes, we have other second-order volitions which interfere with our moral dictates: I simply pushed my Christian training to the background as I satisfied my lust. "I see in my members another law at war with the law of my mind and making me captive to the law of sin which dwells in my members" (Romans 7:23). In the second case, the Christian might talk in terms of third-order volitions about satisfying the dictates of God: I was morally blind, for I disregarded my Saviour's teaching. Or properly regard me as a wanton.

And moreover, through the rocket analogy, we support Augustine's thinking about predestination. There is nothing to stop God – the Werner von Braun of the infinite – from knowing what is going to happen, even though we are homing-device-type rockets rather than shoot-them-off-and-hope-for-the-best-type rockets. We would expect an all-knowing Being to know what is going to happen to us: how we will choose, and what the consequences will be. He is not interfering in our choices. We are free, we are autonomous. Nor is He so designing things that no matter what we do the same end will result. God is not fate. There is therefore space for our own responsibility and our own sin. Where and when and whether we hit the target is a function of outside circumstances and our own adequacy as target searchers – not forgetting built-in gremlins, if you want to push the analogy to the limit. God intended the rockets, but because they are made from material components by human beings, things (as with Apollo 13) are going to go wrong sometimes.

There is also space in all of this for God's judgement, the appropriateness of His righteous punishment, and the need and meaningfulness of His grace and forgiveness. God can even tell who is going to ask for grace and forgiveness. But it is not He who makes us into sheep and goats, even though He knows from the beginning where we will end up in the divine barnyard. And if you are a universalist, believing that eventually we will all be sheep, there is space for this too! Darwinism is ecumenical. Its processes can and will accommodate a wide range of theological options.

Epilogue

Can a Darwinian be a Christian? Absolutely! Is it always easy for a Darwinian to be a Christian? No, but whoever said that the worthwhile things in life are easy? Is the Darwinian obligated to be a Christian? No, but try to be understanding of those who are. Is the Christian obligated to be a Darwinian? No, but realize how much you are going to foreswear if you do not make the effort, and ask yourself seriously (if you reject all forms of evolutionism) whether you are using your God-given talents to the full.

Early in this book, I gave you my pledge that I would try to be honest, not dodging the difficult issues but aiming always to see how a fairly full-blooded version of Darwinism can compare and connect with a fairly traditional and no less full-blooded reading of Christianity. I have tried to stay true to my promise, structuring the discussion by going from re-vealed religion, through natural theology, and on to the nature and foun-dation of moral issues. Some areas, even though they may rate high on the public visibility scale, seem not to be matters of great strain or ten-sion. Obviously, if you are a fundamentalist Christian, then the Darwinian reading of Genesis is going to give you major problems – insoluble prob-lems, I suspect. But, as I have pointed out, biblical interpretation is a topic that Christians have been discussing and refining almost since their religion began. There are plenty of resources open to the Christian who would move towards science and away from a literal reading of the early books of Genesis.

Some areas require still a great deal of thoughtful work and discussion. The notion of original sin, and its origins in the light of Darwinian evolutionary theory, is an issue on which the final word has not yet been spoken. Bound up with this is the whole question of the human soul, its nature, its beginnings, and similar questions. I am speaking now not negatively, but with a sense of incompleteness – and also with a sense of anticipation and excitement. If there is a unifying conclusion to this book it is that while the comparison of Darwinism and Christianity may be challenging and difficult, it is also stimulating and fruitful. I argue – I have argued – that time and again what might seem to be firm barriers to the Darwinian and the Christian existing in one and the same person prove, on examination, to be precisely the points where advances can be made and understandings can be achieved.

Then there is the status of humans and the necessity of their appearance. Contrary to what many think, there is (as there always has been) a strain of the purest Darwinism which encourages precisely the belief that we humans are special, that in some ways (dear to the Christian) we have succeeded in ways that others have not, and that our appearance was not just inexplicable chance. Not all Darwinians accept this thinking, but many do, and paradoxically those who do include people whose concern for the success and well-being of Christianity is minimal, to say the least. There is the question of design and (as we have seen) the related question of pain and suffering. Again Darwinism speaks strongly and positively to the Christian on these matters, throwing significant light on the very existence of those troubling phenomena which apparently testify against an all-loving, all-powerful Deity. If Dawkins is correct, pain and suffering are the necessary cost of getting design (and humans) at all, in any sense.

Then there is the matter of morality. I will simply say what I have said before. I see remarkable parallels between the Darwinian human and the Christian human. On both accounts, there is an internal battle. Human beings are selfish individuals. We look first to our own interests and disregard those of others. We are mean and greedy and aggressive and many other such things. But there is in humankind a sense of concern for our fellow humans, an ethical drive which is not simply one of desire or brute feeling. It is something which involves genuine obligation. We do have real moral feelings for others. Sometimes these sentiments move us into action. Sometimes they do not. But they are there, and that is a shared conviction of science and religion. And finally, there is the matter of freedom and the will. This was the topic of my last chapter, and as you know my conclusion is that, on these matters, there is much in modern

Darwinism which should speak to the most conservative Augustinian thinker. Both Darwinian and Christian are worried about being locked into actions by fate or blind law or something of this nature. And both Darwinian and Christian can find ways forward, showing that the concerns are genuine but that real solutions lie at hand, ready to be taken. Darwinian and Christian have much to learn from each other on this, as on earlier problems.

I draw to an end. I leave it to you to recall other issues, for instance, the collapse of claims that Darwinian naturalism positively refutes or excludes the Christian religion. Is there one last thing to be said? Yes. If you are a Darwinian or a Christian or both, remember that we are mere humans and not God. We are middle-range primates with the adaptations to get down out of the trees, and to live on the plains in social groups. We do not have powers which will necessarily allow us to peer into the ultimate mysteries. If nothing else, these reflections should give us a little modesty about what we can and cannot know, and a little humility before the unknown. Our limitations do not make Christianity mandatory or even plausible, but necessitate a tolerance and appreciation of those who would go beyond science, even if we ourselves cannot follow. Remember Haldane: "I suspect that there are more things in heaven and earth than are dreamed of or can be dreamed of, in any philosophy. That is the reason why I have no philosophy myself, and must be my excuse for dreaming."

Amen!

Bibliography

Allan, J. M. 1869. On the real differences in the minds of men and women. *Journal of the Anthropological Society* 7: cxcv–ccxix.

Alvarez, L. W., W. Alvarez, F. Asaro, and H. V. Michel. 1980. Extraterrestrial cause for the Cretaceous-Tertiary extinction. *Science* 208: 1095–1108.

Appel, T. A. 1987. *The Cuvier-Geoffroy Debate: French Biology in the Decades before Darwin.* New York: Oxford University Press.

Aquinas, St. T. 1963. *Summa Theologiae: 13, Man Made to God's Image (1a. 90–102).* London: Eyre and Spottiswoode.

——— 1966. *Summa Theologiae: 28, Law and Political Theory (1a2ae. 90–97).* London: Eyre and Spottiswoode.

——— 1970. *Summa Theologiae: 11, Man (1a. 75–83).* London: Eyre and Spottiswoode.

Aristotle. 1984. *De Anima.* In Barnes, J., ed., *The Complete Works of Aristotle* (1:641–692). Princeton: Princeton University Press.

Augustine, St. [413–426] 1950. *The City of God,* trans. Dodds, M. New York: Random House.

——— [413–426] 1972. *Concerning the City of God Against the Pagans,* trans. Bettenson, H. Harmondsworth: Penguin.

——— 1982. *The Literal Meaning of Genesis,* trans. Taylor, J. H. New York: Newman.

——— 1983. Commentary on the First Letter of John. In Swift, L. J., ed., *The Early Fathers on War and Military Service* (148). Wilmington, Del.: Michael Glazier.

Ayala, F. J. 1967. Man in evolution: a scientific statement and some theological and ethical implications. *The Thomist* 31(1): 1–20.

——— 1974. Introduction. In Ayala, F., and T. Dobzhansky, eds., *Studies in the Philosophy of Biology* (vii–xvi). Berkeley: University of California Press.

——— 1998. Human nature: one evolutionist's view. In Brown, W. S., N. Murphy, and H. N. Malony, eds., *Whatever Happened to the Soul? Scientific and Theological Portraits of Human Nature* (31–48). Minneapolis: Fortress Press.

Ayer, A. J. 1954. Freedom and necessity. In his *Philosophical Essays* (271–84). London: Macmillan.

Bannister, R. 1979. *Social Darwinism: Science and Myth in Anglo-American Social Thought.* Philadelphia: Temple University Press.

Barbour, I. 1988. Ways of relating science and theology. In Russell, R. J., W. R. Stoeger, and G. V. Coyne, eds., *Physics, Philosophy, and Theology: A Common Quest for Understanding* (21–48). Vatican City: Vatican Observatory.

Barclay, R. 1908. *An Apology for the True Christian Divinity.* Philadelphia: Friends Book Store.

Bartel, D. P., and J. W. Szostak. 1993. Isolation of new ribozymes from a large pool of random sequences. *Science* 261: 1411–18.

Barth, K. 1933. *The Epistle to the Romans.* Oxford: Oxford University Press.

Beach, W., and H. R. Niebuhr, eds. 1955. *Christian Ethics: Sources of the Living Tradition.* New York: Ronald Press.

Behe, M. 1996. *Darwin's Black Box: The Biochemical Challenge to Evolution.* New York: Free Press.

Bergant, D., and C. Stuhlmueller. 1985. Creation according to the Old Testament. In McMullin, E., ed., *Evolution and Creation* (153–75). Notre Dame: University of Notre Dame Press.

Bergson, H. 1907. *L'évolution créatrice.* Paris: Alcan.

Betz, D. 1985. *Essays on the Sermon on the Mount.* Philadelphia: Fortress.

Betzig, L. L., M. Borgerhoff Mulder, and P. W. Turke. 1987. *Human Reproductive Behaviour.* Cambridge: Cambridge University Press.

Bieri, R. 1964. Humanoids on other planets? *American Scientist* 52: 452–8.

Bonhoeffer, D. 1979. *Letters from Prison,* enlarged ed., ed. Bethge, E. New York: Macmillan.

Bowler, P. J. 1976. *Fossils and Progress.* New York: Science History Publications.
    1983. *The Eclipse of Darwinism: Anti-Darwinian Evolution Theories in the Decades around 1900.* Baltimore: Johns Hopkins University Press.
    1984. *Evolution: The History of the Idea.* Berkeley: University of California Press.
    1996. *Life's Splendid Drama.* Chicago: University of Chicago Press.

Bradie, M. 1986. Assessing evolutionary epistemology. *Biology and Philosophy* 1: 401–60.

Brewster, D. 1854. *More Worlds than One: The Creed of the Philosopher and the Hope of the Christian.* London: Camden Hotten.

Buckland, W. 1823. *Reliquiae Diluvianae.* London: John Murray.

Burkhardt, R. W. 1977. *The Spirit of System: Lamarck and Evolutionary Biology.* Cambridge, Mass.: Harvard University Press.

Bury, J. B. [1920] 1924. *The Idea of Progress: An Inquiry into its Origin and Growth.* London: Macmillan.

Cairns-Smith, A. G. 1982. *Genetic Takeover and the Mineral Origins of Life.* Cambridge: Cambridge University Press.
    1985. *Seven Clues to the Origin of Life.* Cambridge: Cambridge University Press.
    1986. *Clay Minerals and the Origin of Life.* Cambridge: Cambridge University Press.

Calvin, J. 1847–1850. *Commentaries on the First Book of Moses Called Genesis,* trans. King, J. Edinburgh: Calvin Translation Society.
    1949. *Institutes of the Christian Religion,* trans. Beveridge, H. London: James Clarke.

1960. *Institutes of the Christian Religion,* ed. McNeill, J. T., trans. Battles, F. L. Philadelphia: The Westminster Press.

Carroll, S. B. 1995. Homeotic genes and the evolution of arthropods. *Nature* 376: 479–85.

Chambers, R. 1844. *Vestiges of the Natural History of Creation.* London: Churchill.

Chomsky, N. 1957. *Syntactic Structures.* The Hague: Mouton.

1966. *Cartesian Linguistics.* New York: Harper and Row.

Conkin, P. K. 1998. *When All the Gods Trembled: Darwinism, Scopes, and American Intellectuals.* Lanham, Md.: Rowman and Littlefield.

Conway Morris, S. 1998. *The Crucible of Creation: The Burgess Shale and the Rise of Animals.* Oxford: Oxford University Press.

Coyne, J. A., N. H. Barton, and M. Turelli. 1997. Perspective: a critique of Sewall Wright's shifting balance theory of evolution. *Evolution* 51(3): 643–71.

Crook, P. 1994. *Darwinism: War and History.* Cambridge: Cambridge University Press.

Cuvier, G. 1813. *Essay on the Theory of the Earth,* trans. Kerr, R. Edinburgh: W. Blackwood.

Daly, M., and M. Wilson. 1988. *Homicide.* New York: De Gruyter.

Darwin, C. 1851a. *A Monograph of the Fossil Lepadidae; or, Pedunculated Cirripedes of Great Britain.* London: Palaeontographical Society.

1851b. *A Monograph of the Sub-Class Cirripedia, with Figures of All the Species. The Lepadidae; or Pedunculated Cirripedes.* London: Ray Society.

1854a. *A Monograph of the Fossil Balanidae and Verrucidae of Great Britain.* London: Palaeontographical Society.

1854b. *A Monograph of the Sub-Class Cirripedia, with Figures of All the Species. The Balanidge (or Sessile Cirripedes); the Verrucidae, and C.* London: Ray Society.

1859. *On the Origin of Species.* London: John Murray.

1871. *The Descent of Man.* London: John Murray.

1872. *The Expression of the Emotions in Man and Animals.* London: John Murray.

Darwin, E. 1803. *The Temple of Nature.* London: J. Johnson.

Darwin, F. ed. 1887. *The Life and Letters of Charles Darwin, Including an Autobiographical Chapter.* London: Murray.

Davies, P. 1999. *The Fifth Miracle: The Search for the Origin and Meaning of Life.* New York: Simon and Schuster.

Dawkins, R. 1976. *The Selfish Gene.* Oxford: Oxford University Press.

1983. Universal Darwinism. In Bendall, D. S., ed., *Molecules to Men* (403–425). Cambridge: University of Cambridge Press.

1986. *The Blind Watchmaker.* New York: Norton.

1995. *A River Out of Eden.* New York: Basic Books.

1997a. Human chauvinism: review of *Full House* by Stephen Jay Gould. *Evolution* 51(3): 1015–20.

1997b. Obscurantism to the rescue. *Quarterly Review of Biology* 72: 397–99.

Dawkins, R., and J. R. Krebs. 1979. Arms races between and within species. *Proceedings of the Royal Society of London, Series B: Biological Sciences* 205: 489–511.

Deacon, T. W. 1997. *The Symbolic Species: The Co-Evolution of Language and the Brain.* New York: Norton.

Degler, C. N. 1991. *In Search of Human Nature: The Decline and Revival of Darwinism in American Social Thought.* New York: Oxford University Press.

Dembski, W. 1998a. *The Design Inference: Eliminating Chance through Small Probabilities.* Cambridge: Cambridge University Press.

——— ed. 1998b. *Mere Creation: Science, Faith and Intelligent Design.* Downers Grove, Ill.: InterVarsity Press.

Dennett, D. C. 1984. *Elbow Room.* Cambridge, Mass.: MIT Press.

——— 1995. *Darwin's Dangerous Idea.* New York: Simon and Schuster.

Depew, D., and B. Weber. 1994. *Darwinism Evolving.* Cambridge, Mass.: MIT Press.

Dick, S. J. 1982. *Plurality of Worlds: The Origins of the Extraterrestrial Life Debate from Democritus to Kant.* Cambridge: Cambridge University Press.

——— 1996. *The Biological Universe: The Twentieth-Century Extraterrestrial Life Debate and the Limits of Science.* Cambridge: Cambridge University Press.

Dillenberger, J. 1960. *Protestant Thought and Natural Science.* London: Collins.

Dobzhansky, T. 1937. *Genetics and the Origin of Species.* New York: Columbia University Press.

Doolittle, R. F. 1997. A delicate balance. *Boston Review* 22(1): 28–29.

Drees, W. 1998. Evolutionary naturalism and religion. In Russell, R. J., W. R. Stoeger, and F. J. Ayala, eds., *Evolutionary and Molecular Biology: Scientific Perspectives on Divine Action* (303–28). Vatican City: Vatican Observatory Press.

Duncan, D., ed. 1908. *Life and Letters of Herbert Spencer.* London: Williams and Norgate.

Ekland, E. H., and D. P. Bartel. 1996. RNA-catalysed RNA polymerization using nucleotide triphosphates. *Nature* 382: 373–6.

Ekland, E. H., J. W. Szostak, and D. P. Bartel. 1995. Structurally complex and highly active RNA ligases derived from random RNA sequences. *Science* 269: 364–70.

Eldredge, N., and S. J. Gould. 1972. Punctuated equilibria: an alternative to phyletic gradualism. In Schopf, T. J. M., ed., *Models in Paleobiology* (82–115). San Francisco: Freeman, Cooper.

Eldredge, N., and I. Tattersall. 1982. *The Myths of Human Evolution.* New York: Columbia University Press.

Ellegård, A. 1958. *Darwin and the General Reader.* Goteborg: Goteborgs Universitets Arsskrift.

Erskine. F. 1995. *The Origin of Species* and the science of female inferiority. In Amigoni, D., and J. Wallace, eds., *Charles Darwin's "The Origin of Species": New Interdisciplinary Essays* (95–121). Manchester: Manchester University Press.

Farley, E., and P. C. Hodgson. 1994. Scripture and tradition. In Hodgson, P. C., and R. H. King, eds., *Christian Theology: An Introduction to Its Traditions and Tasks,* second ed. (61–87). Minneapolis: Fortress Press.

Farley, J. 1977. *The Spontaneous Generation Controversy from Descartes to Oparin.* Baltimore: Johns Hopkins University Press.

Feduccia, A. 1996. *The Origin and Evolution of Birds.* New Haven, Conn.: Yale University Press.

Ferris, J. R., A. R. Hill, R. Liu, and L. E. Orgel. 1996. Synthesis of long prebiotic oligomers on mineral surfaces. *Nature* 381: 59–61.

Feuerbach, L. [1841]1973. *Essence of Christianity*, ed. Schuffenhauer, W. Berlin: Akademie Verlag.

Fisher, R. A. 1930. *The Genetical Theory of Natural Selection*. Oxford: Oxford University Press.

1947. The renaissance of Darwinism. *Listener* 37: 1001.

Fodor, J. 1983. *The Modularity of Mind*. Cambridge, Mass.: MIT Press.

Foote, M. 1990. Nearest-neighbor analysis of trilobite morphospace. *Systematic Zoology* 39: 371–82.

Fox, S. W. 1988. *The Emergence of Life: Darwinian Evolution from the Inside*. New York: Basic Books.

Frankfurt, H. G. 1971. Freedom of the will and the concept of a person. *Journal of Philosophy* 68(1): 5–20.

Frede, M. 1992. On Aristotle's conception of the soul. In M. C. Nussbaum and A. O. Rorty, eds., *Essays on Aristotle's "De Anima"* (93–107). Oxford: Clarendon Press.

Freeman, S., and J. C. Herron. 1998. *Evolutionary Analysis*. Englewood-Cliffs, N.J.: Prentice-Hall.

Friedlander, S. 1997. *Nazi Germany and the Jews: The Years of Persecution 1933–39*. London: Weidenfeld and Nicolson.

Fry, I. 1999. *The Emergence of Life on Earth: A Historical and Scientific Overview*. New Brunswick, N.J.: Rutgers University Press.

Futuyma, D. J., and M. Slatkin, eds. 1983. *Coevolution*. Sunderland, Mass.: Sinauer.

Galileo. 1957. *Discoveries and Opinions of Galileo*, trans. Drake, S. Garden City, N.Y.: Anchor.

Gasman, D. 1971. *The Scientific Origins of National Socialism: Social Darwinism in Ernst Haeckel and the Monist League*. New York: Elsevier.

1998. *Haeckel's Monism and the Birth of Fascist Ideology*. Frankfurt: Peter Lang.

Gilkey, L. B. 1985. *Creationism on Trial: Evolution and God at Little Rock*. Minneapolis: Winston Press.

Gillespie, C. C. 1950. *Genesis and Geology*. Cambridge, Mass.: Harvard University Press.

Ginger, R. 1958. *Six Days or Forever: Tennessee vs. John Thomas Scopes*. New York: Oxford University Press.

Gosse, P. 1857. *Omphalos: An Attempt to Untie the Geological Knot*. London: J. Van Voorst.

Gotthelf, S., and J. G. Lennox, eds. 1987. *Philosophical Issues in Aristotle's Biology*. Cambridge: Cambridge University Press.

Gould, S. J. 1977a. *Ever Since Darwin*. New York: Norton.

1977b. *Ontogeny and Phylogeny*. Cambridge, Mass.: Belknap Press.

1981. *The Mismeasure of Man*. New York: Norton.

1982. Darwinism and the expansion of evolutionary theory. *Science* 216: 380–7.

1988. On replacing the idea of progress with an operational notion of directionality. In Nitecki, M. H., ed., *Evolutionary Progress* (319–38). Chicago: University of Chicago Press.

1989. *Wonderful Life: The Burgess Shale and the Nature of History.* New York: Norton.

1996. *Full House: The Spread of Excellence from Plato to Darwin.* New York: Paragon.

1997a. Darwinian Fundamentalism. *The New York Review of Books* 44 (12 June): 34–37.

1997b. Nonoverlapping magisteria. *Natural History* 106(2): 16–22, 60–2.

1999. *Rocks of Ages: Science and Religion in the Fullness of Life.* New York: Ballantine.

Gould, S. J., and N. Eldredge. 1977. Punctuated equilibria: the tempo and mode of evolution reconsidered. *Paleobiology* 3: 115–51.

Gould, S. J., and R. C. Lewontin. 1979. The spandrels of San Marco and the Panglossian paradigm: a critique of the adaptationist program. *Proceedings of the Royal Society of London, Series B: Biological Sciences* 205: 581–98.

Grant, P. R. 1986. *Ecology and Evolution of Darwin's Finches.* Princeton: Princeton University Press.

Grant, R. B., and P. R. Grant. 1989. *Evolutionary Dynamics of a Natural Population: The Large Cactus Finch of the Galapagos.* Chicago: University of Chicago Press.

Gray, A. 1860. Natural Selection not inconsistent with natural theology. *Atlantic Monthly* 6: 109–116, 229–39, 406–25.

Green, J. B. 1998. "Bodies – that is human lives": a re-examination of human nature in the Bible. In Brown, W. S., N. Murphy, and H. N. Malony, eds., *Whatever Happened to the Soul? Scientific and Theological Portraits of Human Nature* (149–73). Minneapolis: Fortress Press.

Greene, J. C., and M. Ruse. 1996. On the nature of the evolutionary process: the correspondence between Theodosius Dobzhansky and John C. Greene. *Biology and Philosophy* 11: 445–91.

Gunton, C. E., ed. 1997. *The Cambridge Companion to Christian Doctrine.* Cambridge: Cambridge University Press.

Haeckel, E. 1866. *Generelle Morphologie der Organismen.* Berlin: Georg Reimer. 1868. *The History of Creation.* London: Kegan Paul, Trench.

Haldane, J. B. S. 1927. *Possible Worlds, and Other Papers.* London: Chatto and Windus.

1929. The origin of life. In *The Rationalist Annual for the Year 1929* (1–10). London: Watts.

1932. *The Causes of Evolution.* London: Longmans, Green.

Hamilton, W. D. 1964a. The genetical evolution of social behaviour I. *Journal of Theoretical Biology* 7: 1–16.

1964b. The genetical evolution of social behaviour II. *Journal of Theoretical Biology* 7: 17–32.

Haught, J. F. 1995. *Science and Religion: From Conflict to Conversation.* New York: Paulist Press.

Hausfater, G., and S. B. Hrdy, eds. 1984. *Infanticide: Comparative and Evolutionary Perspectives.* New York: Aldine.

Hempel, C. G. 1966. *Philosophy of Natural Science.* Englewood Cliffs, N.J.: Prentice-Hall.

Herschel, J. F. W. 1830. *Preliminary Discourse on the Study of Natural Philosophy.* London: Longman, Rees, Orme, Brown, Green, and Longman.

Hick, J. 1970. *The Philosophy of Religion.* Englewood-Cliffs, N.J.: Prentice-Hall.
1978. *Evil and the God of Love.* New York: Harper and Row.

Hitler, A. 1939. *Mein Kampf.* trans. Murphy, J. London: Hurst and Blackett.

Hodge, C. 1872. *Systematic Theology.* London and Edinburgh: Nelson.

Hodgson, P. C., and R. H. King, eds. 1994. *Christian Theology: An Introduction to Its Traditions and Tasks,* second ed. Minneapolis: Fortress Press.

Holum, J. R. 1987. *Elements of General and Biological Chemistry,* seventh ed. New York: Wiley.

Hopson, J. A. 1977. Relative brain size and behavior in archosaurian reptiles. *Annual Review of Ecology and Systematics* 8: 429–48.

Hrdy, S. B. 1981. *The Woman that Never Evolved.* Cambridge, Mass.: Harvard University Press.

Hume, D. [1779] 1947. *Dialogues Concerning Natural Religion,* ed. Smith, N. K. Indianapolis: Bobbs-Merrill.
1978. *A Treatise of Human Nature.* Oxford : Oxford University Press.

Huxley, J. S. 1927. *Religion without Revelation.* London: Ernest Benn.
1942. *Evolution: The Modern Synthesis.* London: Allen and Unwin.
1948. *UNESCO: Its Purpose and Its Philosophy.* Washington, D.C.: Public Affairs Press.

Huxley, L. 1900. *The Life and Letters of Thomas Henry Huxley.* London: Macmillan.

Huxley, T. H. 1863. *Evidence as to Man's Place in Nature.* London: Williams and Norgate.
1871. Administrative nihilism. In his *Methods and Results* (251–89). London: Macmillan.
1893a. *Darwiniana.* London: Macmillan.
1893b. Evolution and ethics. In his *Evolution and Ethics* (46–116). London: Macmillan.

Isaac, G. 1983. Aspects of human evolution. In Bendall, D. S., ed., *Evolution from Molecules to Men* (509–43). Cambridge: Cambridge University Press.

Jacob, F. 1977. Evolution and tinkering. *Science* 196: 1161–66.

James, K. D., and A. D. Ellington. 1995. The search for missing links between self-replicating nucleic acids and the RNA world. *Origins of Life and Evolution of the Biosphere* 25: 515–30.

Jensen, J. V. 1991. *Thomas Henry Huxley: Communicating for Science.* Newark: University of Delaware Press.

Jerison, H. J. 1973. *Evolution of the Brain and Intelligence.* New York: Academic Press.

Johanson, D., and Edey, M. 1981. *Lucy: The Beginnings of Humankind.* New York: Simon and Schuster.

John Paul II. 1997. The Pope's message on evolution. *Quarterly Review of Biology* 72: 377–83.
1998. *Fides et Ratio: Encyclical Letter of John Paul II to the Catholic Bishops of the World.* Vatican City: L'Osservatore Romano.

Johnson, P. E. 1995. *Reason in the Balance: The Case against Naturalism in Science, Law and Education*. Downers Grove, Ill.: InterVarsity Press.

Jones, G. 1980. *Social Darwinism and English Thought*. Brighton: Harvester.

Joravsky, D. 1970. *The Lysenko Affair*. Cambridge, Mass.: Harvard University Press.

Kant, I. 1929. *Critique of Pure Reason*. trans. Smith, N. K. London: Macmillan.

    1949. *Critique of Practical Reason*, trans. Beck, L. W. Chicago: University of Chicago Press.

    1959. *Foundations of the Metaphysics of Morals*, trans. Beck, L. W. Indianapolis: Bobbs-Merrill.

    [1755] 1981. *Universal Natural History and Theory of the Heavens*, trans. Jaki, S. Edinburgh: Scottish Academic Press.

Kauffman, S. A. 1993. *The Origins of Order: Self-Organization and Selection in Evolution*. Oxford: Oxford University Press.

Kelly, A. 1981. *The Descent of Darwin: The Popularization of Darwinism in Germany, 1860–1914*. Chapel Hill: University of North Carolina Press.

Kepler, J. 1965. *Kepler's Conversation with Galileo's Sidereal Messenger*, trans. Rosen, E. New York and London: Johnson Reprint Corporation.

Kimura, M. 1983. *Neutral Theory of Molecular Evolution*. Cambridge: Cambridge University Press.

Kingsley, F. E., ed. 1895. *Charles Kingsley: His Letters and Memories of His Life*. London: Macmillan.

Korsmeyer, J. D. 1998. *Evolution and Eden: Balancing Original Sin and Contemporary Science*. New York: Paulist Press.

Kropotkin, P. [1902] 1955. *Mutual Aid*, ed. Montague, A. Boston: Extending Horizons Books.

Kuhn, T. 1957. *The Copernican Revolution*. Cambridge, Mass.: Harvard University Press.

Lack, D. 1957. *Evolutionary Theory and Christian Belief: The Unresolved Conflict*. London: Methuen.

Lamarck, J. B. 1809. *Philosophie zoologique*. Paris: Dentu.

Larson, E. J. 1997. *Summer for the Gods: The Scopes Trial and America's Continuing Debate over Science and Religion*. New York: Basic Books.

Levins, R., and R. C. Lewontin. 1985. *The Dialectical Biologist*. Cambridge, Mass.: Harvard University Press.

Lewontin, R. C. 1974. *The Genetic Basis of Evolutionary Change*. New York: Columbia University Press.

    1978. Adaptation. *Scientific American* 239(3): 176–93.

Loewe, L., and S. Scherer. 1997. Mitochondrial Eve: the plot thickens. *Trends in Ecology and Evolution* 12(11): 422–3.

Lucas, J. R. 1979. Wilberforce and Huxley: a legendary encounter. *Historical Journal* 22: 313–30.

Luther, M. 1915. *Works*. Philadelphia: Holman.

    1959. *The Large Catechism*, trans. Fischer, R. H. Philadelphia: Muhlenberg.

    1962. Temporal authority: to what extent it should be obeyed. In Lehman, H. T., ed., *Works*, vol. 45 (75–129). Philadelphia: Muhlenberg.

Mackie, J. 1977. *Ethics*. Harmondsworth: Penguin.

1978. The law of the jungle. *Philosophy* 53: 553–73.

Malthus, T. R. [1826] 1914. *An Essay on the Principle of Population,* sixth edition. London: Everyman.

Marchant, J., ed. 1916. *Alfred Russel Wallace: Letters and Reminiscences.* London: Cassell and Company.

Marsden, G. 1980. *Fundamentalism and American Culture: The Shaping of Twentieth Century Evangelicalism 1870–1925.* Oxford: Oxford University Press.

1990. *Religion and American Culture.* San Diego: Harcourt, Brace, and Jovanovich.

1991. *Understanding Fundamentalism and Evangelicalism.* Grand Rapids, Mich.: Eerdmans.

Marx, K. [1845] 1932. Theses on Feuerbach. In Adoratskii, A., ed., *Marx-Engels Gesamtausgabe,* vol. 1. Berlin: Marx-Engels Verlag.

Maynard Smith, J. 1981. Did Darwin get it right? *London Review of Books* 3(11): 10–11.

1982. *Evolution and the Theory of Games.* Cambridge: Cambridge University Press.

Mayr, E. 1942. *Systematics and the Origin of Species.* New York: Columbia University Press.

1954. Change of genetic environment and evolution. In Huxley, J., A. C. Hardy, and E. B. Ford, eds., *Evolution as a Process* (157–80). London: Allen and Unwin.

McGrath, A. E. 1997. *Christian Theology: An Introduction,* second ed. Oxford: Blackwell.

McMullin, E. 1980. Persons in the universe. *Zygon* 15: 69–89.

1981. How should cosmology relate to theology? In Peacocke, A., ed., *The Sciences and Theology in the Twentieth Century* (17–57). Notre Dame: University of Notre Dame Press.

1993. Evolution and special creation. *Zygon* 28: 299–335.

1996. Evolutionary contingency and cosmic purpose. In Himes, M., and S. Pope, eds., *Finding God in All Things* (140–161). New York: Crossroad.

McNeil, M. 1987. *Under the Banner of Science: Erasmus Darwin and His Age.* Manchester: Manchester University Press.

Medawar, P. B. 1961. Review of *The Phenomenon of Man. Mind* 70: 99–106 .

Meléndez-Hevia, E., T. G. Waddell, and M. Cascante. 1996. The puzzle of the Krebs citric acid cycle: assembling the pieces of chemically feasible reactions, and opportunism in the design of metabolic pathways during evolution. *Journal of Molecular Evolution* 43: 293–303.

Metzger, B. M., and M. D. Coogan, eds. 1993. *The Oxford Companion to the Bible.* Oxford: Oxford University Press.

Miller, K. 1999. *Finding Darwin's God.* New York: Harper and Row.

Miller, S. L. 1953. A production of amino acids under possible primitive Earth conditions. *Science* 117: 528–9.

1992. The prebiotic synthesis of organic compounds as a step toward the origin of life. In Schopf, J. W., ed., *Major Events in the History of Life* (1–28). Boston: Jones and Barlett.

Millhauser, M. 1954. The scriptural geologists: an episode in the history of opinion. *Osiris* 11: 65–86.

Mitman, G. 1990. Evolution as gospel: William Patten, the language of democracy and the Great War. *Isis* 81: 44–93.

Moore, A. 1890. The Christian doctrine of God. In Gore, C., ed., *Lux Mundi* (41–81). London: John Murray.

Moore, G. E. 1903. *Principia Ethica.* Cambridge: Cambridge University Press.

Murphy, N. 1997. *Reconciling Theology and Science: A Radical Reformation Perspective.* Kitchener, Ontario: Pandora Press.

    1998. Human nature: historical, scientific, and religious issues. In Brown, W. S., N. Murphy, and H. N. Malony, *Whatever Happened to the Soul? Scientific and Theological Portraits of Human Nature* (1–29). Minneapolis: Fortress Press.

Nagel, E. 1961. *The Structure of Science: Problems in the Logic of Scientific Explanation.* New York: Harcourt, Brace and World.

Nozick, R. 1981. *Philosophical Explanations.* Cambridge, Mass.: Harvard University Press.

Numbers, R. L. 1992. *The Creationists.* New York: Knopf.

    1998. *Darwinism Comes to America.* Cambridge, Mass.: Harvard University Press.

O'Collins, G. 1993. Resurrection. In McGrath, A. E., ed., *The Blackwell Encyclopedia of Modern Christian Thought* (553–7). Oxford: Blackwell.

Oakes, E. 1998. Original sin: a disputation. *First Things* 87: 16–24.

Oakley, K. P. 1964. The Problem of Man's Antiquity. *Bulletin of the British Museum (Natural History), Geological Series* 9, 5.

Orgel, L. E. 1998. The origin of life – a review of facts and speculations. *Trends in Biochemical Sciences* 23: 491–5.

Oparin, A. 1957. *The Origin of Life on the Earth,* third ed. London: Academic Press.

    [1928] 1967. The origin of life (originally published as *Proishkhozhdenie zhizni*), trans. Synge, A. In Bernal, J. D., ed., *The Origin of Life* (199–234). Cleveland: World.

Osborn, H. F. 1910. *The Age of Mammals in Europe, Asia and North America.* New York: Macmillan.

    1931. *Cope: Master Naturalist: The Life and Writings of Edward Drinker Cope.* Princeton: Princeton University Press.

Palcy, W. [1802] 1819. *Natural Theology* (collected works, vol. 4). London: Rivington.

Pannenberg, W. 1968. *Jesus – God and Man.* London: SCM Press.

Peacocke, A. R. 1986. *God and the New Biology.* London: Dent.

Pelikan, J. 1971–89. *The Christian Tradition: A History of the Development of Doctrine.* Chicago: University of Chicago Press.

Pilbeam, D. 1984. The descent of Hominoids and Hominids. *Scientific American* 250(3): 84–97.

Pinker, S. 1991. Rules of language. *Science* 253: 530–35.

    1994. *The Language Instinct: How the Mind Creates Language.* New York: William Morrow.

    1997. *How the Mind Works.* New York: Norton.

Pittenger, M. 1993. *American Socialists and Evolutionary Thought, 1870–1920.* Madison: University of Wisconsin Press.

Pius XI. 1933. *On Christian Marriage.* London: Sheed and Ward.

Plantinga, A. 1991a. An evolutionary argument against naturalism. *Logos* 12: 27–49.

   1991b. When faith and reason clash: evolution and the Bible. *Christian Scholar's Review* 21(1): 8–32.

   1993. *Warrant and Proper Function.* New York: Oxford University Press.

   1997. Methodological naturalism. *Perspectives on Science and Christian Faith* 49(3): 143–54.

Polkinghorne, J. 1989. *Science and Providence: God's Interaction with the World.* Boston: Shambhala.

   1994. *Science and Christian Belief: Theological Reflections of a Bottom-Up Thinker.* London: SPCK.

Popper, K. R. 1974. Darwinism as a metaphysical research programme. In Schilpp, P. A., ed., *The Philosophy of Karl Popper,* vol. 1 (133–43). LaSalle, Ill.: Open Court.

Popper, K. R., and J. Eccles. 1977. *The Self and Its Brain.* Berlin: Springer International.

Powell, B. 1855. *Essays on the Spirit of the Inductive Philosophy.* London: Longman, Brown, Green, and Longmans.

   1860. On the study of the evidences of Christianity. In *Essays and Reviews* (94–144). London: Longman, Green, Longman, and Roberts.

Provine, W. B. 1971. *The Origins of Theoretical Population Genetics.* Chicago: University of Chicago Press.

   1988. Progress in evolution and meaning in life. In Nitecki, M. N., ed., *Evolutionary Progress* (49–74). Chicago: University of Chicago Press.

Puccetti, R. 1968. *Persons: A Study of Possible Moral Agents in the Universe.* London: Macmillan.

Quine, W. V. O. 1969. *Ontological Relativity and Other Essays.* New York: Columbia University Press.

Quinn, P. L. 1978. *Divine Commands and Moral Requirements.* Oxford: Clarendon Press.

Ramsey, P. 1950. *Basic Christian Ethics.* New York: Charles Scribner's Sons.

Rawls, J. 1971. *A Theory of Justice.* Cambridge, Mass.: Harvard University Press.

Reichenbach, B. R. 1982. *Evil and a Good God.* New York: Fordham University Press.

Richards, R. J. 1987. *Darwin and the Emergence of Evolutionary Theories of Mind and Behavior.* Chicago: University of Chicago Press.

Ruse, M. 1973. *The Philosophy of Biology.* London: Hutchinson.

   1975a. Charles Darwin's theory of evolution: an analysis. *Journal of the History of Biology* 8: 219–41.

   1975b. The relationship between science and religion in Britain, 1830–1870. *Church History* 44: 505–22.

   1975c. Darwin's debt to philosophy: an examination of the influence of the philosophical ideas of John F. W. Herschel and William Whewell on the development of Charles Darwin's theory of evolution. *Studies in History and Philosophy of Science* 6: 159–81.

   1979a. *The Darwinian Revolution: Science Red in Tooth and Claw.* Chicago: University of Chicago Press.

1979b. *Sociobiology: Sense or Nonsense?* Dordrecht: Reidel.

1980. Charles Darwin and group selection. *Annals of Science* 37: 615–30.

1982. *Darwinism Defended: A Guide to the Evolutionary Controversies.* Reading, Mass.: Addison-Wesley.

1984. Is there a limit to our knowledge of evolution? *BioScience* 34(2): 100–104.

1986a. *Taking Darwin Seriously: A Naturalistic Approach to Philosophy.* Oxford: Blackwell.

1986b. Evolutionary ethics: a phoenix arisen. *Zygon* 21: 95–112.

ed. 1988a. *But Is It Science? The Philosophical Question in the Creation/Evolution Controversy.* Buffalo, N.Y.: Prometheus.

1988b. *Homosexuality: A Philosophical Inquiry.* Oxford: Blackwell.

1989. *The Darwinian Paradigm: Essays on Its History, Philosophy and Religious Implications.* London: Routledge.

1993. Evolution and progress. *Trends in Ecology and Evolution* 8(2): 55–59.

1995. *Evolutionary Naturalism: Selected Essays.* London: Routledge.

1996a. *Monad to Man: The Concept of Progress in Evolutionary Biology.* Cambridge, Mass.: Harvard University Press.

1996b. The Darwin industry: a guide. *Victorian Studies* 39(2): 217–35.

1999. *Mystery of Mysteries: Is Evolution a Social Construction?* Cambridge, Mass.: Harvard University Press.

Russell, E. S. 1916. *Form and Function: A Contribution to the History of Animal Morphology.* London: John Murray.

Russell, R. J. 1998. Special providence and genetic mutation: a new defense of theistic evolution. In Russell, R. J., W. R. Stoeger, and F. J. Ayala, eds., *Evolution and Molecular Biology: Scientific Perspectives on Divine Action* (191–223). Vatican City: Vatican Observatory Publications.

Russett, C. E. 1976. *Darwin in America: The Intellectual Response, 1865–1912.* San Francisco: Freeman.

Schleiermacher, F. 1928. *The Christian Faith.* Edinburgh: T. and T. Clark.

Schwartz, J. 1991. Reduction, elimination, and the mental. *Philosophy of Science* 58: 203–20.

Scott, E. C. 1996. Monkey business: creationists regroup to expel evolution from the classroom. *The Sciences* (January/February): 20–25.

1997. Antievolutionism and Creationism in the United States. *Annual Review of Anthropology* 26: 263–89.

Settle, M. L. 1972. *The Scopes Trial: The State of Tennessee v. John Thomas Scopes.* New York: Franklin Watts.

Simpson, G. G. 1944. *Tempo and Mode in Evolution.* New York: Columbia University Press.

1949. *The Meaning of Evolution.* New Haven, Conn.: Yale University Press.

1964. *This View of Life.* New York: Harcourt, Brace, and World.

Singer, P. 1981. *The Expanding Circle: Ethics and Sociobiology.* New York: Farrar, Straus, and Giroux.

Skyrms, B. 1996. *Evolution of the Social Contract.* Cambridge: Cambridge University Press.

Spencer, H. 1851. *Social Statics; or the Conditions Essential to Human Happiness Specified and the First of Them Developed.* London: J. Chapman.

  1852a. A theory of population, deduced from the general law of animal fertility. *Westminster Review* 1: 468–501.

  1852b. The development hypothesis. In his *Essays: Scientific, Political and Speculative* (377–83). London: Williams and Norgate.

  1857. Progress: its law and cause. *Westminster Review* 67: 244–67.

  1862. *First Principles.* London: Williams and Norgate.

  1864. *The Principles of Biology.* London: Williams and Norgate.

  1892. *The Principles of Ethics.* London: Williams and Norgate.

  1904. *Autobiography.* London: Williams and Norgate.

Stanley, S. M. 1979. *Macroevolution, Pattern and Process.* San Francisco: W. H. Freeman.

Stebbins, G. L. 1950. *Variation and Evolution in Plants.* New York: Columbia University Press.

Sumner, W. G. 1914. *The Challenge of Facts and Other Essays.* New Haven: Yale University Press.

Swinburne, R. G. 1970. *The Concept of Miracle.* London: Macmillan.

Taylor, P. W., ed. 1978. *Problems of Moral Philosophy.* Belmont, Calif.: Wadsworth.

Teilhard de Chardin, P. 1955. *Le phénomène humaine.* Paris: Editions de Seuil.

Temple, F. 1884. *The Relations between Religion and Science.* London: Macmillan.

Trivers, R. L. 1971. The evolution of reciprocal altruism. *Quarterly Review of Biology* 46: 35–57.

Van Inwagen, P. 1983. *An Essay on Free Will.* Oxford: Oxford University Press.

Vermeij, G. J. 1987. *Evolution and Escalation: An Ecological History of Life.* Princeton: Princton University Press.

Wallace, A. R. 1900. *Studies: Scientific and Social.* London: Macmillan.

  1903. *Man's Place in the Universe.* London: Chapman and Hall.

  1905. *My Life; A Record of Events and Opinions.* London: Chapman and Hall.

Wallwork, E. 1982. Thou shalt love thy neighbour as thyself: the Freudian critique. *Journal of Religious Ethics* 10: 264–319.

Ward, K. 1996. *God, Chance and Necessity.* Oxford: Oneworld.

  1998. *God, Faith and the New Millennium.* Oxford: Oneworld.

Watson, J. D., and F. H. C. Crick. 1953. Molecular structure of nucleic acids. *Nature* 171: 737.

Weldon, W. F. R. 1898. Presidential address to the zoological section of the British Association. *British Association for the Advancement of Science: Report of the Sixty-Eighth Meeting, Bristol, September 1898* (887–902). London: John Murray.

Wesley, J. n.d. The use of money. In his *Sermons on Several Occasions.* London: Methodist Publishing House.

Westman, R. S. 1986. The Copernicans and the churches. In Lindberg, D. C., and R. L. Numbers, eds., *God and Nature: Historical Essays on the Encounter between Christianity and Science* (76–113). Berkeley: University of California Press.

Whewell, W. 1837. *The History of the Inductive Sciences.* London: Parker.

1840. *The Philosophy of the Inductive Sciences.* London: Parker.

1853. *The Plurality of Worlds.* London: Parker.

Whitcomb, J. C., and H. M. Morris. 1961. *The Genesis Flood: The Biblical Record and Its Scientific Implications.* Philadelphia: Presbyterian and Reformed Publishing Company.

Wilson, E. O. 1975. *Sociobiology: The New Synthesis.* Cambridge, Mass.: Harvard University Press.

1978. *On Human Nature.* Cambridge, Mass.: Harvard University Press.

1992. *The Diversity of Life.* Cambridge, Mass.: Harvard University Press.

1994. *Naturalist.* Washington, D.C.: Island Books/Shearwater Books.

Wittgenstein, L. 1980. *Culture and Value,* ed. Von Wright, G. H., trans. Winch, P. Oxford: Blackwell.

Wogaman, J. P. 1993. *Christian Ethics: A Historical Introduction.* Louisville: Westminster/John Knox Press.

Wolpoff, M., and R. Caspari. 1997. *Race and Human Evolution.* Boulder: Westview.

Wooldridge, D. 1963. *The Machinery of the Brain.* New York: McGraw-Hill.

Wright, S. 1931. Evolution in Mendelian populations. *Genetics* 16: 97–159.

1932. The roles of mutation, inbreeding, crossbreeding and selection in evolution. *Proceedings of the Sixth International Congress of Genetics* 1: 356–366.

Young, R. M. 1985. *Darwin's Metaphor: Nature's Place in Victorian Culture.* Cambridge: Cambridge University Press.

Zwingli, H. [1526] 1995. This is my body. In McGrath, A., ed., *The Christian Theology Reader* (308–10). Oxford: Blackwell.

Index

Darwin —
God as designer through
unspoken law (not miracle)

"Sometimes you just hang in there and
hope everybody dies."

→ on debate between cultures
politics too